An Introduction to Keeping Sheep

An Introduction to Keeping Sheep

Second Edition

JANE UPTON

and

DENNIS SODEN

Farming Press

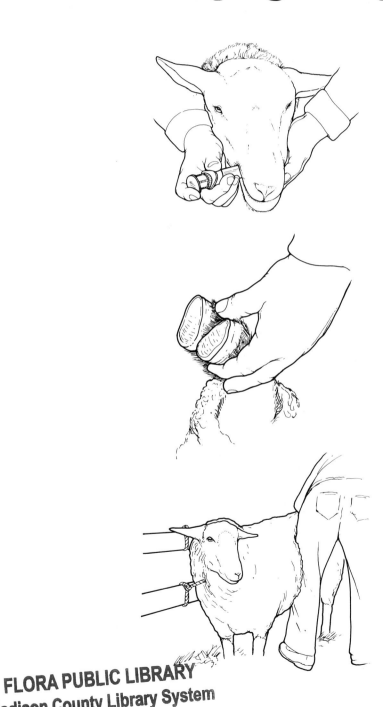

Acknowledgements

Our thanks go to **Gill McLellan** for her expert typing and practical knowledge of sheep-keeping that assisted in the compilation of this book; to **Mr A.J. Stevens MA, BVSC, Dip Bact, MRCVS** for his encouragement and advice on some of the diseases affecting sheep; to **Hilary Howell** who kindly guided us through the maze of legislation affecting the welfare aspects of sheep husbandry; to **our families** for their continuing forbearance in the way that we have disrupted the domestic scene in our homes; to **our friends** and also **members of the Small Shepherds Club. Thank you all!**

First published 1991
Second edition 1996
Reprinted with amendments 1999

ISBN 0 85236 332 X

A catalogue record for this book is
available from the British Library.

**Published by Farming Press,
Miller Freeman UK Ltd,
2 Wharfedale Road,
Ipswich IP1 4LG,
United Kingdom**

*Distributed in North America by
Diamond Farm Enterprises,
Box 537, Bailey Settlement Road,
Alexandria Bay, NY 13607, USA*

Photographs and drawings by Jane Upton
Cover design by Andrew Thistlethwaite
Typeset by Winsor Clark
Printed and bound in Great Britain by Butler and Tanner, Frome, Somerset

Contents

Foreword   vii

INTRODUCTION   1

1 GETTING STARTED   2

2 WHAT TYPE OF SHEEP TO BUY   4

3 PLANNING YOUR SHEEP AREA   22

4 LEARNING HOW TO HANDLE YOUR SHEEP   30

5 CARING FOR YOUR SHEEP   36

6 CARING FOR THE BREEDING FLOCK   48

7 LAMBING   54

8 SOME COMMON DISEASES AFFECTING SHEEP   74

9 KEEPING RECORDS   86

10 THE SHEPHERD'S CALENDAR   93

11 SHOWING   95

12 GRASSLAND MANAGEMENT   101

13 MISCELLANEOUS INFORMATION CONCERNING SHEEP   103

INDEX   109

Foreword to the Second Edition

The authors have added much new material to this second edition, thus making the book even more well rounded — one which I confidently recommend to all keepers of sheep.

Let us not forget that the United Kingdom is in the forefront of the European sheep industry in the production and marketing of meat and wool, and is already expanding into the milk production sector.

We keep sheep for four main reasons: to produce meat, wool and milk and for decorative reasons, e.g. park sheep. Depending upon one's environment and desires, one can choose 'horses for courses'. The new section on selecting your breed is very relevant and very important.

Three or more millenniums ago the Mediterranean countries were developing their sheep industry as one of their basic resources, using the Merino and fat-tailed desert sheep. We also read constantly of sheep throughout the Bible. In the north our forebears were relying more on the small St Kilda/Soay ancestors and mountain (wet weather) sheep. In comparatively modern times the Southdown and Leicester breeders were developing many of the breeds we now know. In the early 19th century men like MacArthur took the Merino strains to Australia and founded a huge industry of enormous financial influence, also crossing them to produce the Corriedale, Polworth and others.

In our industry we are constantly developing and expanding, shedding the inefficient, retaining rare breeds — in case we need their genes — and always probing. In the middle of this century Oscar Colburn, a great thinker, evolved the Colbred. We are seeing the Suffolks developing into a new highly specialised and well-recorded sheep breed.

The Dorset Horn has become to a large extent a polled breed. The Dorset Down, once popular, is now evolving after 170 years into a leaner, more commercially viable animal. Consider too, the Jacob — how it has pulled itself together in 25 years of intensive work by very keen breeders.

We no longer have those very canny traditional shepherds with whom my generation grew up. The lessons we learned from them would fill another book. You who read, learn and act on this book are going to be responsible for forming part of the next chapter in the sheep's history. You are an integral part of our ever-changing scene. Hone your skills, use your intelligence and be part of this great industry. The basics are all here in this book.

Sheep have played a major part in my life; their calming yet gregarious characteristics have a great levelling effect upon the daily chores and worries of our lives. I have found sheep people to be different from other keepers of livestock. Consider all these factors, together with all the practical chapters of this book — and then take your place in the order of things.

C.G. HYDE
Wool, Dorset
February 1996

An Introduction to Keeping Sheep

Introduction

Those of us who have watched a group of lambs playing in the spring sunshine or seen the miracle of a new-born lamb taking its first breath will readily agree that there is a great deal of pleasure to be experienced in keeping and caring for a flock of sheep.

This book is written from a background of many years of practical experience in large and small flocks of both pedigree and commercial sheep, and it seeks to help and guide those who are keeping sheep for the first time. It describes in detail how and why various practical tasks are carried out and how to avoid some of the pitfalls and awkward situations that can arise from time to time. It describes some of the differences between different types of sheep, and it suggests how, when and where to set about starting a flock.

CHAPTER 1 Getting Started

What has prompted your decision to keep sheep? Are they to act purely as grass mowers with the added bonus of a homegrown, maybe organically produced, supply of meat? Does the prospect of keeping breeding animals appeal? Is there a need to diversify in an existing livestock enterprise?

In any event the sheep will need an area of land on which they can graze. That land will have to be securely fenced, with a supply of water nearby. The prospective owner will have decided whether sufficient spare time is available to care for the sheep throughout the year, and whether help will be available for periods when holidays or sickleave are taken. The amount of time required to give adequate care to the sheep will vary from two short visits per day in the summer after shearing, to frequent visits during each twenty-four hour period during lambing time.

Once these basic requirements have been considered and the decision has been taken to go ahead, it is time to start making plans.

Making contacts

Sheep keepers come from all walks of life, and there might be a local club where people who keep sheep as a hobby can get together and discuss experiences and organise talks, visits and demonstrations. Your local vet should be able to tell you of others who keep small flocks. Visit local sheep fairs and auction sales and meet and talk with as many sheep keepers as possible.

Assessing the available land

The size of your flock will to a large extent be dictated by the amount of land that is available for grazing. Assess the quality and quantity of the herbage growing on it, remembering that you will expect your sheep to live almost entirely from it for up to eight months of the year. Is it 'light' or 'heavy' land? Heavy land is close-textured, usually clay-based and is generally more productive than light land, but it will not stand heavy stocking during periods of high rainfall as it poaches or gets trampled into mud very quickly. Light land is composed of larger particles of soil, is free draining and does not poach so easily, but the herbage growing on it does suffer during periods of hot weather and drought. Remember also when you are assessing the grazing that your flock may increase by as much as two hundred per cent following a very successful lambing.

How many sheep?

Initially stocking rates should be no more than five ewes per hectare. This number can be adjusted in subsequent years if necessary. The sheep will be healthier and more contented if your land is understocked rather than overstocked.

Where to find your sheep

If there is a sheep or shepherd's club operating locally this can be a good source

A sheep auction

as members often have surplus stock for sale. Look at the small ads in your local paper, or your vet may know of sheep for sale. Be wary of small groups being offered for sale in the local auction market. They may be rejects from a large flock for any number of reasons including the ability to escape from any type of enclosure! However, a trip to a local auction market or sheep fair can be useful as a guide to the value of different classes of sheep and it will give you an idea of how much you must expect to pay when you make your purchase.

If you decide that you would like to keep a pure breed of sheep you will be able to obtain the name and address of the secretary of the breed society from the National Sheep Association, The Sheep Centre, Malvern, Worcestershire WR13 6PH. The secretary will be able to let you know where you can obtain specimens of your preferred breed, and whether any are available in your locality. If one of the rare breeds has taken your fancy, details can be obtained from the Rare Breeds Survival Trust, National Agricultural Centre, Stoneleigh, Warwickshire CV8 2LG.

CHAPTER 2 **What Type of Sheep to Buy**

The type of sheep to buy will be determined to a certain extent by what is available and the time of year when you decide to purchase. Most of the sheep fairs and sales of breeding sheep take place between August and November, but sheep from private sources are available at all times of the year. Ewes of four or five years of age that have already produced lambs are probably the best buy for beginners as they will be experienced in motherhood. Another possibility is to buy weaned ewe lambs in the autumn but not to mate them until the autumn of the following year. This will give the beginner an opportunity to get to know the sheep and how to handle them. Ewe lambs are much lighter in weight and the various tasks are more easily carried out with them than with heavier sheep. You will become more proficient in handling as their weight increases!

If the sheep are required purely to keep grass under control, then consider buying wether lambs, i.e. castrated males, as their feed requirements are less than for breeding ewes. They are usually sold by their breeders in the autumn and fattened during the next few months before being sold for meat. If they are kept as grass mowers then they will need to be shorn and dipped in the spring and will have increased considerably in size and weight by the following autumn. They can then be sold for meat and replacement lambs purchased. The main drawback to this is that they will be offered for sale at a time when smaller, younger lambs are also being sold for meat, and the price per kilo tends to decrease as the weight and age of the sheep increase. Yet another possibility is to buy wether lambs of a longwoolled or finewoolled breed and keep them for a number of years, using the income from their high-value fleeces as a contribution towards the cost of keeping them.

Conformation

There are certain basic standards which should be looked for:
● The sheep should have a good mouth with incisor teeth which meet the pad on the top jaw in the correct position — neither overshot nor undershot.

NORMAL JAW

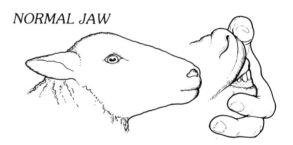

UNDERSHOT JAW

OVERSHOT JAW

GOOD OVERALL CONFORMATION

- There should be good strong bone and straight legs, not bowing inwards or outwards. Pay particular attention to the hind legs of rams. Ensure that there are no weaknesses in the pastern, which is the joint immediately above the foot. Any weakness there can affect the performance of rams and ewes kept for breeding.
- A long level back is desirable with a good spring of rib giving plenty of room for the heart and lungs.

*STRONG BONE
&
STRAIGHT LEGS*

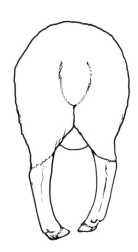

A WEAK PASTERN

*LEGS SHOULD NOT BE BOWING
OUTWARDS OR INWARDS*

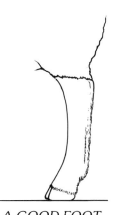

A GOOD FOOT

WHICH BREED TO BUY

There are over eighty different breeds of sheep in the United Kingdom, including recent imports from overseas, all trying to find a niche in the British sheep industry. The British breeds can be categorised into three main types: Shortwools, Longwools and Hill or Mountain.

The Shortwools

These include the Down breeds which were developed in the southern half of England by using a Southdown ram on ewes of various local breeds.

The Southdown was fixed as a breed and generally improved for meat production by selection from within the breed by John Ellman at Glynde near Lewes in Sussex during the 18th century. At the beginning of the 20th century huge flocks — by today's standards — were kept on large estates, producing lamb and mutton of the first quality. By the end of the Second World War numbers had fallen dramatically as meat rationing and the general food shortage gave quantity precedence over quality, and the breed has never recovered its former popularity.

Another reason for the decline of the breed was the shortage of shepherds willing or qualified to work with such labour-intensive flocks. In recent years Southdowns have been imported from New Zealand in an attempt to make the breed more acceptable to modern conditions. They are longer-bodied and longer in the leg than their original UK forebears, without losing the quality for which the breed is famed.

The Southdown was used to develop the other Down breeds by crossing with local breeds to produce the Hampshire Down, the Dorset Down, the Suffolk and the Oxford Down in various parts of the country.

Because of their docility and close flocking instinct, Down breeds were used to improve the fertility of the soil through a system known as folding. The sheep were confined by a temporary fence of hurdles or wire netting to a small area of a crop grown specially for them. Each day they were moved on to a fresh area so that their dung was distributed evenly over the field. This benefited subsequent grain crops and at the time was one of the principal sources of fertility, especially on outlying areas of the farm.

The Down breeds are now almost entirely used to produce terminal sires for crossing with the Masham, Mule, Scottish and Welsh Halfbreds and similar crossbred ewes where the progeny are destined for the meat market.

The Hampshire Down   The origin of this breed stems from crosses between the Southdown and the Wiltshire Horn and the now extinct Berkshire Knot. Centred on Hampshire, Wiltshire and Berkshire, it is a breed which at the present time is used mainly as a terminal sire of very early-maturing butcher's lambs. Many pedigree flocks are Meat and Livestock Commission recorded and have their lambs backfat scanned and so breed from lean, fast-growing sheep.

The Dorset Down   This medium sized, brownfaced breed evolved in the parish of Wool, in Dorset, in the 1820s, and is the result of using a Southdown ram on Hampshire and Wiltshire ewes. Dorset ewes will take the ram at most times of the year. The lambs are very early maturing, whether from purebred stock or from crossbred ewes, and regularly give a carcase of 18kg in 12 weeks. The wool quality is among the highest of the native British breeds and considerable quantities go to the hosiery trade.

Southdown

Hampshire Down

Dorset Down

Oxford Down

The Oxford Down   This is the largest of the Down breeds, and became established in the 1830s. The parentage consisted of Cotswold rams and Hampshire and Southdown ewes.

One of the main attributes of the breed is its ability to sire lambs which have the potential to grow to heavy weights without excess fat. The carcases are ideal for boning and rolling as they are large with a good depth of lean meat. Mature rams of this breed can weigh over 150kg.

Suffolk

Polled Dorset

The Suffolk This is the main terminal breed sire in the UK being used in 70 per cent of prime lamb-producing crossbred flocks. It has a distinctive clean black head and legs and was first recognised as a pure breed in 1810. It evolved from crossing Southdown rams with Norfolk Horn ewes which ran on the heathlands of East Anglia. There are over 2000 pedigree Suffolk breeders in the UK, distributed all over the country, and ram sales are held countrywide. Suffolk cross lambs are early maturing, reaching slaughter weight in 12 to 14 weeks; they can grow on to heavier weights later in the year without producing excess fat. Many producers use the Suffolk sired crossbred ewe lambs as commercial dams in their flocks, producing prime lambs, particularly the Suffolk X Scotch Halfbred and the Suffolk X Mule and Masham.

The Shropshire Following the establishment of the breed in the 19th century, the Shropshire reached its peak of popularity early in the 20th. It went into decline in the 1920s and it was not until the 1980s that it began to make a recovery. Today it is a sheep well-suited to present-day requirements, whether it is bred pure or used as a crossing sire.

The Dorset Horn and its sister breed the **Polled Dorset**, which was developed in Australia, share an important characteristic in that they will breed at almost any time of the year. In some cases it is possible to breed three crops of lambs in two years. The ewes are exceptionally good milkers which explains why their progeny can make high-quality carcases at ten weeks of age.

The Ryeland is a whitefaced breed rather similar in appearance to the Southdown. It is a breed of great antiquity and its origins in the county of Hereford go back 800 years. Its wool is of high quality and as a breed it is very docile.

The Clun Forest is another breed which has declined in popularity since the years just after the Second World War, when it was in great demand. It originates from south-west Shropshire adjoining the Welsh Border and it is a very adaptable breed, being found in most parts of the UK. The ewe is a prolific breeder and produces excellent prime lambs when crossed with Down breed sires. When mated to the Border Leicester she produces the increasingly popular English Halfbred, which is in demand as a commercial ewe.

The Kerry Hill This black and white, strikingly handsome breed from the Kerry Hills of Montgomeryshire probably had a common origin with the Welsh Mountain sheep and was recognised as a distinct breed before the beginning of the 19th century. Although not numerically strong, it is particularly suited to ley farming. The Kerry Hill ewe is prolific and a good milker, producing good-quality prime lambs when crossed with Down breed rams.

There are several breeds from Wales and the Welsh borders which represent the transition from the breeds of the mountains to those found on the lower altitude of the uplands and into the lowlands. These include the Hill Radnor, the Beulah and the Lleyn.

The Hill Radnor is equally at home on the hills, uplands and lowlands of the Welsh Marches, where it was developed many years ago. It was bred pure with the ewe lambs for replacements on the hills and the wether lambs either sold for meat straight off their dams or sold to lowland farmers for further fattening. It has a tan face and legs and is extremely hardy, docile and under favourable conditions it is a prolific breeder with adequate milk for its lambs.

The Beulah Speckled Face This breed is bigger and less hardy than true mountain sheep and because of a management system where they spend part of the year confined in fields they are much more docile. It is a very handsome breed with a black and white face and a fleece of clean, soft, white wool. As a pure breed the ewes can produce butcher's lambs of 36kg liveweight in 12 to 15 weeks. In recent years the Border Leicester and the Bluefaced Leicester rams have been used to provide Welsh Mule ewe lambs for commercial flocks.

The Lleyn Lleyn sheep take their name from the Lleyn peninsula in North Wales.

They are medium-sized, polled and whitefaced. Although their numbers declined in postwar years, they were rescued by a few devotees of the breed and are now firmly established again. High prolificacy, strong mothering instinct and high milk yield enable it to produce prime carcases of around 20kg from grass.

The Longwools

Generally speaking, the Longwools are larger sheep which mature more slowly than the Shortwools. They were more popular when families were larger and consequently larger joints of mutton were required. Their fleeces provided an income from the relatively poorer hill areas of the Cotswolds and the Yorkshire and Lincolnshire Wolds. These areas are the homes of the Cotswold, the Leicester and the Lincolnshire Longwool breeds. The Longwool group also includes the Romney Marsh, which originates from the low-lying area of that name in Kent.

The Romney has the ability to withstand very high stocking densities and can carry a higher worm burden than other breeds without showing ill-effect. It has been exported successfully to many parts of the world and on to all kinds of terrain. When crossed with the Southdown ram in New Zealand, it was responsible for the famous Canterbury lamb. The Romney is still the most popular breed in the area which bears its name.

Some of the Longwools, notably the Border Leicester, the Teeswater, the Wensleydale and the Hexham or Bluefaced Leicester, are used almost exclusively to produce rams for crossing with ewes of the hill breeds. The resulting ewe lambs from this cross are sold to lowland farmers, either as lambs or as 18 month old ewes, where they are mated with Shortwool rams to produce prime lamb for the meat market.

9

The **Teeswater** is from the Northern Pennines where, for at least 150 years, it has been bred by farmers to provide rams for crossing with Hill ewes, mostly Swaledale, Dalesbred, Rough Fell and Herdwick. The resulting cross is known as the Masham — a highly prolific sheep much in demand by commercial prime lamb producers. The Teeswater is a medium-sized, hornless breed with a long, lustrous fleece. It has been developed over the years on purely commercial lines and it is judged by the qualities of its Masham progeny. Fancy points are largely ignored and the ram is rarely used for pure breeding until it has proved itself as a breeder of good-quality Mashams. The Masham wether lamb can, under a variety of conditions, produce a long, lean, prime carcase ideally suited to today's needs. The Teeswater is an exceptionally prolific, heavy-milking ewe, and lambing averages of 250 to 300 per cent are not uncommon.

The **Border Leicester** is a breed established in Scotland some 250 years ago, and results from crosses between the Cheviot and English Leicester breeds. It is a large, white-faced and hornless breed, and is used principally to produce high-quality female progeny when crossed with the smaller Hill breeds. When crossed with the Cheviot the progeny are known as the Scottish Halfbred, with the Blackface it is known as the Scottish Greyface, and with the Welsh Mountain it produces the Welsh Halfbred. All of these crossbred ewes are used by commercial farmers to breed prime-quality lamb when put to terminal sires of the Down breeds.

The **Hexham or Bluefaced Leicester**
The importance of this breed lies in the qualities it imparts to its crossbred offspring. It is a large sheep and the rams are used chiefly on the Swaledale, Dalesbred, Rough Fell and Herdwick ewes, the lambs being known as the North of England Mule.

Border Leicester

Further descriptions of the qualities attributed to the Mule and Masham crossbred ewes are to be found on pages 12-13 under the Swaledale and Dalesbred headings.

The **Wensleydale**   This is a large, bluefaced sheep with a long, lustrous fleece. Both ewes and rams will produce up to 7kg of wool with a staple length of 30cm. Its main use is for crossing with the Hill breeds to produce commercial ewes which are prolific and heavy-milking and which, when mated with terminal sires, are producers of prime-quality lambs.

Wensleydale

Other members of this group include the Dartmoor, the Devon and Cornwall Longwool and the Devon Closewool, which come from the high rainfall areas of south-west England. Mature rams of the Devon and Cornwall breed can produce around 16kg of wool annually, as can rams of the Lincoln, Leicester and Cotswold breeds.

The Devon Closewool North Devon, where this breed originated and from which it has spread over large areas of south-west England, is one of the most densely sheep-populated areas of the world. Rarely seen outside the south-west, the Devon Closewool is medium-sized with a white face and produces a good fleece of 3 to 4kg of medium-quality wool used for hosiery and tweeds. Many flocks live entirely off grass all the year round.

The Dartmoor The **Greyfaced Dartmoor** is a sheep whose fleece and constitution have been developed to cope with the extreme climatic conditions of the bleak moorland which is its home. Numerically it is not as strong as many other breeds. The wool is long in staple and lustrous, and fleeces from yearling rams average 12kg in weight with an individual fibre length of 22cm. The **Whitefaced Dartmoor,** another breed found on the moor, has, as its name implies, a white face. It is very hardy, being kept out on the moor from May to December. Formerly all of the breeds were horned but now the majority, particularly of the ewes, are polled.

The Cotswold is a large whitefaced sheep of great antiquity. A Cotswold shearling ram will produce a fleece of up to 12kg in weight, while fleeces from shearling ewes range from 5.5 to 6.75kg. In the Middle Ages Cotswold sheep were the chief source of the wealth which enabled wool merchants to build many of the attractive stone manor houses and churches seen in the Cotswold Hills today. The breed was used in the early

Devon & Cornwall Longwool

Dartmoor

19th century to cross with Hampshire Down ewes to produce the Oxford Down.

In the 1960s the breed became almost extinct, the total ewe flock being confined to just one owner. From the dispersal of this flock, a number of small flocks were started and since then numbers have grown. A flock book has been started and the future of this famous breed is now assured.

11

The Hill Breeds

The hill breeds occupy a unique position in the sheep industry in that their distribution covers a huge area of high land which is largely unsuitable for any other type of farming. Two qualities which all the hill breeds possess are hardiness and thriftiness. Hardiness gives them the ability to withstand severe climatic conditions, and thriftiness the ability to thrive on a limited food supply, often of poor quality.

They are smaller sheep than either of the two previous categories and, unlike the Down breeds, they have evolved a habit of scattered grazing. They spread themselves thinly on the ground so that there is less competition for the available grazing. When enclosed in a lowland situation they regard a field boundary as a challenge to see what lies beyond, as they believe the grass is always greener on the other side! This behaviour is understandable when it is appreciated that they are the descendants of animals which have been given free range over a huge area of countryside for many hundreds of years.

The Blackface   The Blackface is numerous in Scotland, Ireland, the north of England and the moors of south-west England. It is a hardy, horned breed with a fleece that helps it to withstand the heavy rainfall, high winds and harsh winters of these areas. There are several different types within the breed which have evolved to match the environmental conditions. The Perth type is large-framed with a medium to heavy fleece and is found mainly in north-east Scotland, south-west England and Ireland. The Northumberland has a softer fleece and is used to produce the North of England Mule when mated with the Hexham Leicester. The Lanark type, which has combined with the Newton Stewart and covers much of Scotland and Ireland, has a shorter fleece and an improved milking ability over other types of Blackface.

The breed is bred pure for four or five

Blackface

years to provide flock replacements, and they are then sold on to kinder land and mated with either the Border Leicester to produce Greyface ewe lambs, or the Hexham or Bluefaced Leicester. Both crosses provide prolificacy and good mothering qualities for the production of prime butcher's lambs when mated to terminal sires.

Blackface rams intended for shows and sales are sometimes dipped in a solution containing an orange/yellow dye in an attempt to enhance their appearance. This is called bloom dipping and can be seen on the ram illustrated. Bloom dipped fleeces purchased by the British Wool Marketing Board attract a penalty.

The Swaledale is probably the strongest breed of sheep numerically, and it is found mainly in the north of England. It is a very hardy breed with a distinctly marked face — the upper part dark and the lower part white or mealy. It has a fleece which is softer than could reasonably be expected from a hardy hill breed. Except at the back of the head, where it is mixed with black, the wool is white with a thick, deep bed and curly top. The Swaledale ewes, which are bred pure initially, are sold at around five years of age and crossed with Teeswater or Bluefaced Leicester rams for the production of much-sought-after ewe lambs. The cross from the Teeswater is known as the Masham, and from the Bluefaced Leicester the North of England Mule. These crosses are to be found in all areas of the UK, where prolificacy, excellent mothering qualities and the heavy milk yields combine with the early maturity from the Down breed sires to give top-quality butcher's lambs.

The Dalesbred   As its name implies, the Dalesbred, which is a cousin of the Swaledale, was developed in the Western Dales of Yorkshire and adjoining areas. It was fixed as a distinct breed in 1930. It has a dark face, a mealy nose and a white mark on each side of its nostrils. Its wool is softer than the Swaledale's, with a dense undercoat and a curly top with no black fibres. As a breed it is used to graze the high ground and to produce three or four crops of purebred ewes, after which the ewes are sold to go on to lower ground where they are crossed with Teeswater and Hexham Leicester rams to provide prolific, milky female progeny, which are again sold on as the dams of prime lambs when crossed with Down breeds.

The Rough Fell is a hardy, long-lived breed which is well fitted to endure the rigours of the high mountains and valleys of the Lake District and the Yorkshire Dales. Both sexes are horned and have a long, coarse fleece which is used for carpets and mattresses. The ewes breed purebred lambs for three or four crops before being sold to farms on kinder land for a further three or four crops of crossbred lambs, the females from which go on to be the dams of prime lambs.

The Herdwick holds pride of place for hardiness and can exist on the sparse and poor-quality grazing of the Lake District and High Fells under wintry conditions, often without any supplementary feeding. It is a

Rough Fell

white-faced breed and only the rams have horns. The lambs are born with black fleeces but as mature sheep the wool fades to a blue-grey colour. The ewes are allowed to mature before being mated, and shearling ewes are not always put to the ram. A mature ewe weighs from 35 to 45kg, and under hill conditions will produce a 90 per cent lamb crop. When taken from the fells and put onto a reasonable plane of nutrition, they are more prolific and will produce a good butcher's lamb to a terminal sire.

Herdwick

13

The Lonk is one of the largest of the native hill breeds. It hails from the Lancashire and Yorkshire Pennines, and can be traced back to 1740. The face is black and white and both sexes are horned. The fleece is of a higher quality than is normally carried by hill breeds. The wool is white and free of kemp. Lonk rams are occasionally used to introduce prolificacy and size into other hill breeds without sacrificing the hardiness for which they are renowned.

The Exmoor Horn is a hill breed which does equally well on moorland grazing. It is whitefaced with horns, and carries a white fleece of medium length and good quality. The draft ewes are in demand for crossing with the Bluefaced Leicester to produce the Exmoor Mule for lowland areas.

The Derbyshire Gritstone A black- and whitefaced hill sheep from the Peak District of Derbyshire, the breed is polled in both sexes. The Gritstone is hardy and thrifty and many flocks spend their entire lives on the hills. When bred pure it is capable of producing prime finished lamb off unimproved land up to 2000 feet above sea level. It crosses well onto the other hill breeds, improving size, carcase quality and wool.

The Cheviot A whitefaced sheep of the Scottish Borders. When bred pure it can produce a quality lamb of 17kg, but more usually the wether lambs are overwintered and fed on turnips on lower land to reach a carcase weight of around 21kg. The draft ewes from the hill were originally crossed with the Border Leicester to produce the famous Scottish Halfbred. Now the Bluefaced Leicester is also used to produce the Cheviot Mule, the females from both crosses making excellent mothers of quality lamb for the meat market. Cheviot rams have been used in Wales on the Welsh Mountain ewe to produce the Brecknock

Cheviot which is now fixed as a separate breed.

The North Country Cheviot originated as the Cheviot in the Border country between England and Scotland. In the 18th century a large number were driven north to occupy land depopulated by the Highland Clearances. Occasional crossing of the Cheviot with the Leicester breed led to a larger-bodied sheep, which is now firmly established and can be seen in large numbers in the most northerly counties of Scotland.

The Welsh Mountain comes in a number of different types. It varies between the small, extremely hardy sheep of the high mountains to the larger type of the common land around the valleys of South Wales. All have white or tan faces and are extremely thrifty, being active foragers.

After a spell of pure breeding, the draft ewe crosses well with the Border Leicester and the Bluefaced Leicester, producing the Welsh Halfbred and the Welsh Mule. Many Welsh ewes are bred to Down rams, and the mothering qualities and the milking ability of the ewe are such that comparatively young lambs grow to a size greater than their dams. Welsh Mountain lamb and mutton is lean, tender and unsurpassed for flavour and comes in the small joints which are popular today, the carcases finishing at from 11 to 18kg.

One feature of the Welsh Mountain is that it is a very active sheep and as such is difficult to confine, especially when brought onto low ground. Any kind of fence will be closely inspected and if at all possible an escape route through, under or over will be found. This is understandable when one realises that in the high mountains the sheep's very survival depended on its being able to seek out meagre grazing ahead of its flock mates. In consequence, as a breed, it was necessary to spread itself thinly over the ground.

14

There are a number of variations of the Welsh Mountain. Some are the result of a recessive gene occurring in the traditional flock. The **Black Welsh Mountain** with its all-black face and wool; the **Torddu or Badgerfaced** with black eye stripes and black belly; and the **Torwen** which has the reverse colouring, white belly and white face markings on a dark coloured body — all have similar characteristics to their white ancestors. They lamb easily, are hardy and thrifty and are attractive in appearance.

Black Welsh Mountain

Badgerfaced

Torwen

There are some other British breeds which do not fit neatly into any of the three main categories discussed. These include the Wiltshire Horn, the Jacob and the Shetland.

The Wiltshire Horn   Unique among British breeds of sheep, the Wiltshire Horn does not carry any wool. It is a lowland breed, an excellent converter of grass, and the breed and its crosses can reach carcase weights of 20kg off grass with little or no concentrates. There are no shearing costs and fly and tick problems are virtually unknown. It is an excellent prime lamb sire, and the rams are extremely active and prepotent in that they transmit liveliness to their lambs at birth and achieve early maturity. The ewes are very protective mothers and the absence of wool means that the newborn lamb can find the teats easily. At birth the lamb has a covering of fine hair which protects it from the cold, and the fine, narrow head and neck make for easy lambing.

The Jacob   The Jacob is a very popular sheep with small flock owners as it is a prolific, commercially viable breed which is instantly recognisable by its black and white patterned fleece. It comes in polled, two-horned and multi-horned varieties and its devotees claim a high resistance to foot problems. The high quality wool in two distinct colours enables interestingly patterned garments to be made from natural, undyed yarns. It meets a ready demand from home spinners.

The Shetland   The Shetland is one of the smallest breeds of British sheep. It is noted for the fineness and softness of its wool and the flavour of its meat. Only the rams are horned and the fleeces come in several shades of colours — white, grey, moorit (reddish brown), fawn and brown.

Recently an attempt has been made to produce a fine-woolled sheep for the hills which will give a heavier fleece and a ewe

Wiltshire Horn

Jacob

16

Shetland

which will give a prime lamb without the necessity for crossing. The new breed is called the **Lomond** and comes from crosses between the Shetland and the English Merino.

RARE BREEDS

These are mainly British breeds that have lost their popularity over the years for various reasons, as fashions come and go in the sheep world and as the demand for meat and wool changes. Many breeds which were popular earlier this century faced extinction after the Second World War. Many traditional breeds were developed by farmers to meet the particular requirement of a small local area, but with a wider dissemination of knowledge and an efficient transport system, sheep could be taken far from their home base to perform as well as, if not better than, they did at home. These rare breeds are now being saved from extinction mainly by interested hobby farmers and rare breed establishments.

In order to rescue some of the breeds,

and to preserve their genes for possible future use, the Rare Breeds Survival Trust was formed. The contributions received from members maintain registration facilities for all the breeds which do not have breed societies. The Trust also keeps a directory and can put prospective members in touch with owners of the breed selected.

The rare breeds include a number of multihorned breeds — the black **Hebridean** and the brown **Manx Loghtan.** The **North Ronaldsay** from the Orkney Islands has adapted its diet so that it can live on seaweed, and the **Boreray** exists mostly as a feral flock on the island of that name in the St. Kilda group.

The Portland   The Portland is an attractive, small, tan-faced sheep, originating in south-west England. Together with the Merino it is believed to be an ancestor of today's Dorset Horn. As a breed it is not prolific, very few of the ewes producing twins, but it does produce a lean, lightweight carcase of great flavour.

Portland

Manx Loghtan

The Norfolk Horn   This breed was rescued from the brink of extinction in the mid 1950s, and, thanks to a grading-up programme using Suffolks, is now once again firmly established. It is a heathland sheep from East Anglia, blackfaced, with both sexes horned. Its cross with the Southdown in the early 18th century produced the modern Suffolk.

Llanwenog

Soay

The Llanwenog   The Llanwenog was fixed as a distinct breed in the relative isolation of the Teifi valley in the late 19th century. Imported Shropshire rams were used on a local horned, bluefaced sheep and it has now developed into a dark-faced grassland sheep producing prime meat lambs. It is very prolific, and lambing percentages of over 200 per cent have been recorded by the Meat & Livestock Commission. It crosses well with all the terminal sires and also with the traditional sires of crossbred ewes retained for commercial prime lamb production.

The Soay   The Soay bears a close resemblance to the wild mouflon sheep and all the present Soay stock is descended from the flock on the island of Soay in the St. Kilda group. It is a brownfaced sheep with either fawn or chocolate coloured wool. Mature ewes weigh around 25kg. Lambing difficulties are rare, and the lambs are very active as soon as they are born. Both sexes are horned.

18

BREEDS FROM ABROAD

In recent years many breeds have been imported into the UK from various parts of the world.

The Texel One of the earliest imports was the Texel, brought in from Holland and France. It was imported to improve the conformation and carcase quality of the native breeds, particularly leanness and in the leg of mutton (gigot). The Texel figures prominently among the prizewinners of carcase competitions, whether purebred or crossed.

Bleu du Maine

Texel

The Bleu du Maine Also from France comes the Bleu du Maine, a large breed with a slate blue face and legs. It has made headway not just as a terminal sire where its progeny can be taken to all weights without getting too fat, but also as the sire of crossbred ewes from upland flocks, giving improved conformation and proflicacy in their offspring.

Charollais

The Charollais This is another breed making a name for itself as a very good carcase sheep. It comes from France and is a medium-sized, heavy sheep. It ranks highly among terminal sires through the identification of superior genetic material within the breed, and the best are used extensively in leading flocks.

The Rouge de l'Ouest   Another terminal sire from France which has performed consistently well in performance testing trials at Loughgall in Northern Ireland. It is a breed which is fertile all the year round and it can also pass this fertility on to its crossbred progeny, an important factor for those who wish to provide out-of-season lamb without recourse to scientific interference.

Rouge de l'Oest

The Vendeen, Cotentin and **Berrichon du Cher**   These breeds, all from France, expect to give increased returns through their crossbred progeny by improving carcase quality.

The Est à Laine Merino   This is a large whitefaced breed which breeds naturally all the year round without the use of hormone implants and injections. Its Merino ancestry means that it carries a fleece of very fine wool, a characteristic which is passed on to its crossbred progeny, giving greatly increased returns from the sale of fleeces off the ewes kept in commercial flocks.

The Polwarth   The main claim to fame of this breed is in the quality of its wool. It is slower maturing but this enables a wool clip to be taken before the lambs are sold, and it has a lean carcase at slaughter. The Polwarth originates from Australia and has been imported into this country from Tasmania. The very fine wool, which comes from its Merino ancestry, is used in the production of high quality worsted cloths. Mature ewes will shear 6 to 7kg of wool and lambs 3 to 4kg at six months. Recently this wool has been making £3 per kg.

The Merino originated in Spain. In spite of a monopoly imposed by the Spanish king, specimens of the breed were smuggled into other parts of Europe in the 18th century. Its fine wool was a unique and profitable crop, so although it had no pretensions as a meat breed, it spread rapidly throughout Europe.

By early in the 19th century the Merino had arrived in Australia, and the era of fine

Merino

wool production in the southern hemisphere had begun. Since then the breed has been developed and used in the formation of many different breeds in practically all parts of the world. Today a Merino ram produces around 10kg of wool annually and a ewe 8kg.

New Zealand Southdown   See the Southdown breed description earlier in this chapter.

These must appear as a bewildering number of breeds for the beginner who is trying to decide which to keep. In making a decision there are a number of factors which need to be considered. Some breeds are more docile than others, and some are endowed with a wanderlust which makes them difficult to enclose in small fields. Body size and weight vary enormously, and this fact should be borne in mind by those who might have difficulty in handling the sheep. Home spinners may be attracted by a particular wool quality or colour, whilst others might go

New Zealand Southdown

for the visual attraction of horns, face colour or markings.

There is a breed to satisfy just about every requirement!

CHAPTER 3

Planning Your Sheep Area

FENCING

Before purchasing your sheep, make sure that the enclosure is securely fenced. If the sheep are allowed to escape they can be a constant source of annoyance to yourself and to your neighbours. The fence can be a natural hedge but unless it is one that has recently been laid the sheep will browse on it and eventually create weak spots. Avoid poisonous species such as yew, rhododendron etc. to which they might be attracted at certain times of the year, particularly when grass is in short supply. Brambles can become entangled in the wool so that the sheep is unable to free itself, and the bark can be stripped from hedgerow and cultivated species in the spring when the sap is rising, thus ring-barking and killing trees and bushes.

Far superior is well-erected fence consisting of a stock fence with a code number of C8/80/15. This is of rectangular mesh wire, with the mesh graduated so that the smallest mesh is near the ground so as to prevent the sheep putting their heads through and either getting stuck or exerting undue pressure so as to weaken the fence. Avoid using any stock fencing with a larger mesh size than C8/80/15.

This netting is 800mm high and additional height can be obtained by adding strands of plain or barbed wire above it. A strand of barbed wire below the fence at ground level will prevent stock from creeping underneath and will also discourage dogs and predators coming in from the opposite direction.

Any wire fence is only as sound as its

Stock fencing

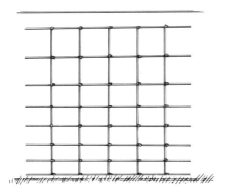

*STOCK FENCE C8/80/15
WITH ADDED PLAIN WIRE
FOR ADDITIONAL HEIGHT AND
BARBED WIRE AT GROUND LEVEL*

construction, and it is vitally important that the fence is correctly tensioned and the supporting stakes are not more than two metres apart. The strainer posts at either end of a straight run of fencing should be at least 900mm into the ground, tightly rammed and well strutted. Somewhere on the fence, either at a strainer post or anywhere along its length, fit a means of tightening the wire when it becomes slack through age or frequent atmospheric temperature changes. Eye bolts, winders and radisseurs can be used as tensioners. When erecting the fence do not drive the holding staples into the intermediate stakes beyond the point where they just touch the wire. This will allow the wire to move through the staples when tensioning or when expansion or contraction occurs. The staples should be at a slight angle when driven into the stakes so that they enter different grains in the wood and avoid splitting.

If the fence is to reinforce a hedge, it should be sited a few feet away from the hedge. This keeps the sheep from climbing onto the fence to feed on the growing tips of the hedge plants and subsequently weakening it. Such behaviour can be discouraged by fitting a single line of electric fence wire so that it projects at right angles

A strainer post

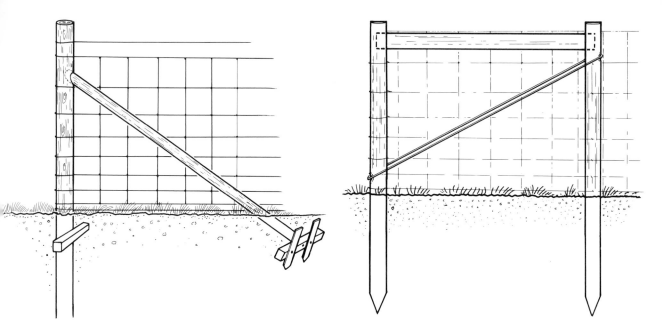

TWO TYPES OF STRAINER POSTS WITH STRUTS

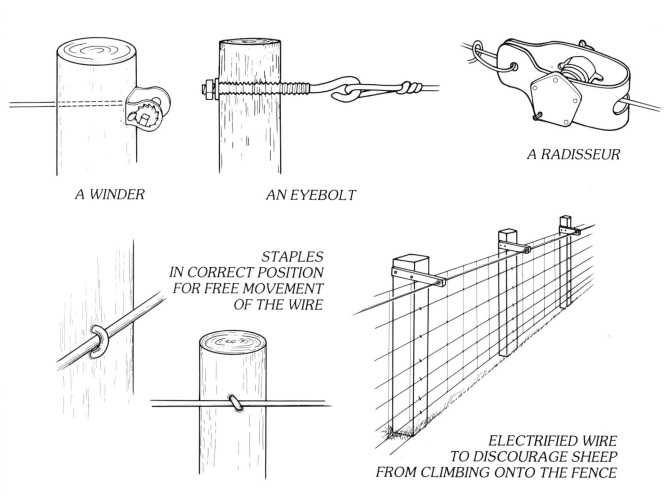

A WINDER

AN EYEBOLT

A RADISSEUR

STAPLES
IN CORRECT POSITION
FOR FREE MOVEMENT
OF THE WIRE

ELECTRIFIED WIRE
TO DISCOURAGE SHEEP
FROM CLIMBING ONTO THE FENCE

24

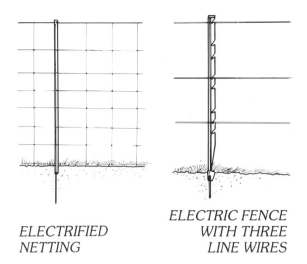

ELECTRIFIED NETTING

ELECTRIC FENCE WITH THREE LINE WIRES

from the intermediate fence posts about 150mm from the top of the wire netting.

Electric fencing can be used as an alternative to the stock fencing. It can be in the form of netting or single line wires. The netting consists of nylon mesh with fine lightweight wire twisted into it to carry the electric current. Single strands of the same materials are also available for use with insulators on wooden posts or on metal posts with built-in insulators. More permanent electric fences using No. 12 galvanised steel wire with insulators on metal or wooden posts does not suffer damage from rabbits and hares as the nylon netting sometimes does. The number of line wires (two or three) and the distance between the wires and the total height of the fence will vary according to the breed.

It is advisable to use an electric fence unit connected via a transformer to the mains electricity supply, or if a portable source of electricity is required use a battery operated fencer unit. Batteries have a habit of running down at critical times, weekends, holidays, etc. The best way to make sure a battery does not run out is to have two, one in the field and the other at home being topped up by a trickle charger. Care should be taken at all times to prevent the electric fencing from

becoming ineffective through earthing by allowing herbage to grow on to it, or by allowing it to come into contact with metal which is not insulated. Very young lambs do not respect electric fencing in the same way as adult sheep. As their small hooves make very light contact with the ground they receive no shock from the fence, and they can become entangled in the electrified netting. However, electric fencing is a good option when the grazing is rented rather than owned, as it can be taken away if use of the land is lost, and put up somewhere else at no further expense. Portable electric fencing is also very useful for making temporary divisions in a field.

If possible site any gate in the boundary fence near a corner of the grazing area. This will make it easier when moving the sheep. It will also be an advantage if the gate opens through 180 degrees; some of the sheep, especially young lambs, will inevitably go to the wrong side if it is at 90 degrees. If the entrance to the grazing is to be used frequently it is advisable to raise the level of the gateway a few inches above the surrounding land with hard core or scalpings so that it remains free from mud even in the worst weather conditions.

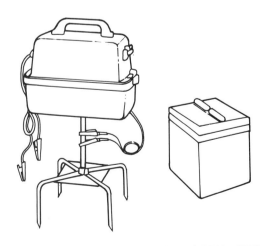

A BATTERY-OPERATED ELECTRIC FENCER UNIT

A SHEEP HOUSE

HOUSING

The grass on the grazing area makes little or no growth through the winter months and both sheep and land will benefit from being rested. During this time the sheep will need supplementary feed and it makes it more comfortable for sheep and shepherd if the flock can be kept under cover for a few weeks. This is also a convenient way of buying extra acres if the food consumed at this time has been grown elsewhere.

The sheep house need not be an elaborate building, and it is sufficient to use a simple pole barn construction. Posts which have been pressure-treated with preservative e.g. secondhand telegraph poles, will last much longer in the ground than untreated ones. A mono-pitch roof will be sufficient and the building should be sited with its back to the prevailing wind with a good overhang to the roof both front and back. If the front overhang is angled downwards, it will prevent rain from blowing onto the bedding.

Barley straw provides the best bedding material. It is superior to wheat, rye or oat straw, all of which very soon get trampled flat whereas barley straw stays light and open textured for much longer. The straw should be clean, bright and free from mould. The ewes will pick it over, eating most of the 'flag' or leaf from the stems. Soiled bedding should be cleared from the house periodically to prevent the build-up of moist heat which can lead to foot problems.

Allow four square metres for each ewe. This will also allow you to accommodate the lambs if necessary. The building should have solid walls to the back and ends to a height of 1200mm, with Yorkshire cladding, consisting of narrow vertical boards with a 20mm gap between them, above. Heavyweight perforated plastic can also be used but this is difficult to tension. Such construction will allow adequate ventilation with no draught at floor level.

The fence at the front can be combined with a feed barrier which will enable the sheep to be fed without entering the pen. There must be sufficient space for all the sheep including shy feeders to feed comfortably. A minimum width of 500mm per animal is recommended. The entrance to the house should be a gate not less than 2m in width. The ideal floor is made from rammed chalk, but hardcore covered with smaller material will also be satisfactory. The height of the finished floor should be above the level of the surrounding land so as to avoid problems with drainage. Solid concrete floors should be avoided as the bedding quickly becomes damp and respiratory problems can arise. You may prefer to feed your sheep on a hardstanding in front of the building. Again, the feed can be placed in the troughs without interference from the sheep and you will be able to inspect them more easily.

Water should be available at all times in small valve-operated troughs with provision

26

made for any water resulting from leakages or blocked valves being directed outside the building. If there is a supply of water available from a conventional cattle drinking trough, provide a step or build up the surrounding area so that the sheep can drink comfortably. When young lambs are present, float a heavy plank of wood on the surface of the water. This will carry the weight of any lamb jumping onto the trough when playing and save it from drowning. It will also prevent the growth of algae in the trough during warm weather.

HANDLING PENS

When the building is no longer required for holding sheep, it can also be used as a handling pen. First clear all the soiled litter from the floor and give a liberal dressing of lime. For small numbers of sheep the only other equipment needed is two or three hurdles hinged together with string and fixed at one end to a side wall of the house. These can be opened out straight and then closed around the sheep, holding them in a confined space while dosing, vaccinating, etc.

If no housing is available then a small, securely fenced enclosure should be constructed in a corner of the field, preferably near the entrance and exit gate. It should consist of a pen large enough to hold all your sheep together, with a small area for individual handling. It is also advisable to build a funnel approach to the pen as this will make penning so much easier. A piece of hessian or similar material stretched between two people can be useful when driving sheep into a narrowing entrance to the pen. If it is found necessary to have a hard floor to the outdoor handling pen, this can be made from loose pebbles around 25mm to 50mm in diameter. These will revolve as the sheep walk on them and are self-cleaning. A retaining board around the outside of the pen will be required to prevent the pebbles from spreading.

The gates to the outdoor handling pen should be strongly constructed and hung so that they close onto the sheep as they are penned. The fastenings should be about 500mm from the ground as this is where almost all of the pressure exerted by a pen

A sheep house doubling as a handling pen

Enticing sheep into the handling area

filled with sheep will come. It is a great help to be able to step over the internal fences of the outdoor handling pen and a suggested height for these is 900mm. They should be of solid construction with occasional 'foot-holes' built in to assist those who are unable to step over. The perimeter fences need to be a little higher at 1,200mm in order to contain the more athletic breeds.

Two things to bear in mind are that

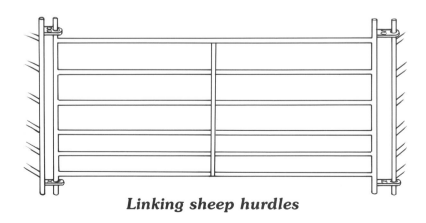

Linking sheep hurdles

sheep will move uphill more freely than downhill, and towards light rather than towards dark, solid, high walls or buildings. You may also consider it worthwhile having a shallow tray or footbath where sheep requiring treatment for footrot or scald can be stood in a solution of zinc sulphate for a while.

If the sheep are kept on rented land where it is not possible to build a permanent handling area, a temporary one can be made using metal hurdles, securely held together. These are easily moved, and can be taken to wherever the sheep are, so that there is always an area in which they can be caught and held for inspection or treatment.

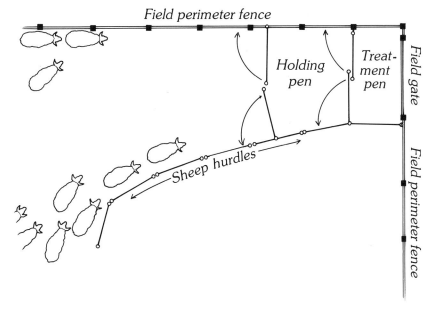

A TEMPORARY HANDLING PEN WITH FUNNEL SHAPED APPROACH

CHAPTER 4 **Learning How to Handle Your Sheep**

During the course of the year it is necessary to handle your sheep for dosing, vaccinating, foot trimming, etc. As described in the previous chapter, confine them in a small area so that the handler does not have to put on a rodeo show each time one needs to be caught! Do not be tempted to grab the wool as the sheep finds this very painful. In any event there is no wool on a newly shorn sheep.

CATCHING

There are three ways in which a sheep can be caught and restrained:

1. Firmly grasp a back leg just above the hock and lift the leg off the ground.
2. Grasp the loose skin in the flank of the sheep just in front of a back leg and lift the leg from the ground.
3. Catch the lower jaw of the sheep in a cupped hand and lift the head above its natural horizontal position. Place the fingers of the other hand under the tail. It can now

be moved backwards or forwards, but generally it is easier to move a sheep backwards, keeping the head up all the time.

If you wish to examine the teeth or read an ear number, it will be sufficient to back the sheep into a corner while still keeping control of the head, but you may find it easier to do this when the sheep is completely immobilised.

IMMOBILISING

There are several ways of immobilising a sheep, and the following directions apply to right-handed people.

It can be done by sitting the sheep on its hindquarters and holding it between the knees, leaving both hands free for the examination. This will also be the position required when paring overgrown feet or treating footrot. To achieve this, having got the sheep into position with its chin held by a cupped hand, and standing on its left side, move the left hand around the neck so that

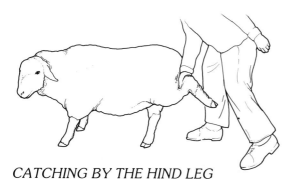

CATCHING BY THE HIND LEG

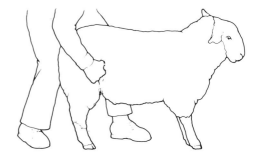

GRASPING LOOSE SKIN IN THE FLANK

30

with the left hand cupped under its chin and place the right hand on its right hip bone. Move the sheep close to the legs, with the left knee behind its left shoulder and the right knee in front of its left hip, and pull its head round so that it is looking back over its right shoulder. At the same time press firmly downwards with the right hand and move the right leg smartly away from contact with the sheep and it will collapse onto its hindquarter.

With experience this method of immobilising a sheep requires very little effort, but once the sheep is in position it is

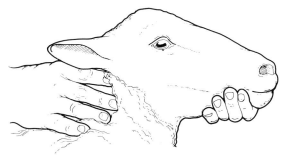

CATCHING THE LOWER JAW OF THE SHEEP IN A CUPPED HAND

Moving the sheep backwards or forwards

the head is resting on the wrist and pull the sheep towards you. Place the left knee just behind the sheep's left shoulder and the right knee just in front of the sheep's left hip. Reach over and firmly grasp the loose skin in front of the sheep's right hind leg. Lean backwards and by using the knees as a pivot the sheep will roll onto its side on the knees of the operator. It can be lowered gently onto either its left or its right hindquarter. Avoid trying to make the sheep sit vertically on its tail as it finds this very uncomfortable.

Another method of sitting the sheep onto its hindquarters is to hold the sheep

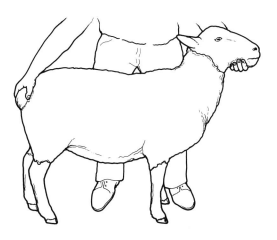

MOVING THE SHEEP BACKWARDS OR FORWARDS

31

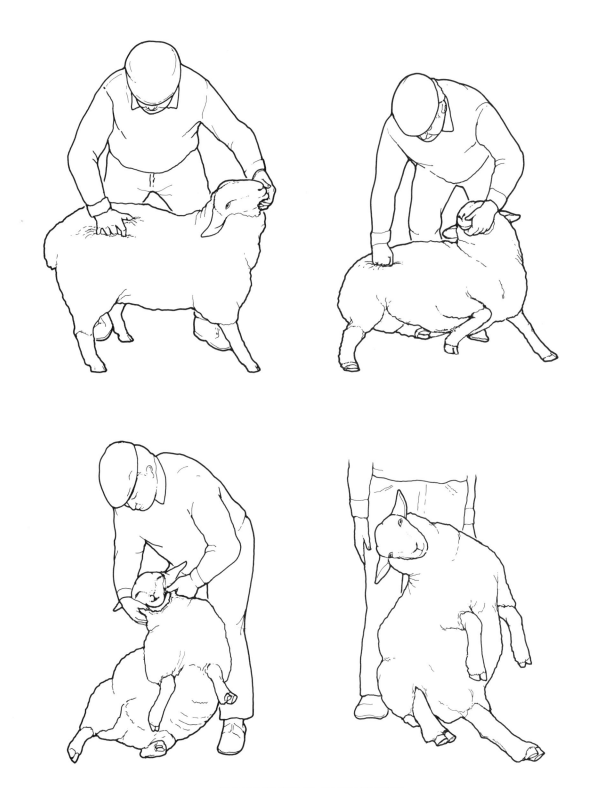

IMMOBILISING THE SHEEP

IMMOBILISING THE SHEEP BY ROLLING IT ONTO THE KNEES OF THE OPERATOR

Immobilising

necessary to prevent its feet from coming in contact with the ground or fences etc. because if they do the sheep will kick and struggle as it attempts to regain the standing position.

There are a number of pieces of equipment on the market which can be used to immobilise sheep. The simplest of these is similar to a stretcher which can be placed at an angle against a fence. The animal is moved backwards until its rear end touches the frame and then the head is lifted until eventually the sheep is lying on its back on the stretcher. A rope is tied across the sheep just below its front legs and the operator is then free to use both hands to treat the sheep, as when trimming feet.

Other items include a sheep 'cradle' which uses a series of clamps and levers to secure the sheep and to move it into an upside down position. This is helpful for anyone who has a back problem, but, as with the previous item, it has the disadvantage to the manual method in that the sheep has to be caught and moved to the equipment before it can be treated.

For heavily pregnant ewes who need assistance at lambing time, another method of immobilising is described on page 61.

It is possible to restrain sheep simply by passing the head through a loop of rope fixed to the side of the handling pen and placing the left knee just in front of the animal's left hip while pushing it against the side of the pen. This can be useful while dagging, crutching or treating blowfly strike in the tail area.

To examine the teeth, use the right hand to hold up the head and part the lips, using the thumb and index finger of the left hand to expose both the upper fibrous pad and the incisor teeth in the lower jaw. There are no incisor teeth in the upper jaw.

A SHEEP STRETCHER

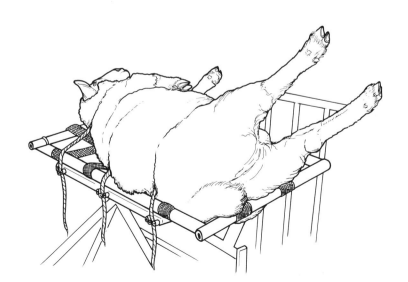

*A MECHANICAL DEVICE
FOR IMMOBILISING
THE SHEEP*

35

CHAPTER 5

Caring for Your Sheep

Adequate food and water are two essentials to the wellbeing of your sheep. The actual amount consumed will vary according to the type of sheep kept, and also from season to season, but try to maintain good bodily condition at all times. Be prepared to offer supplementary food when the nutritional requirements of the sheep are not being met, because of poor quality or insufficient grazing. The condition required on any particular type of sheep can only be learned by experience, and a visit to a local market with a knowledgeable companion will be useful.

Supplementary Food

The type of supplementary food offered will depend on the time of year when the grass is in short supply and also on the type of

sheep. Good-quality hay made from grass cut at the early flowering stage, which has been baled and carted in dry weather and has no trace of mould, will usually provide for maintenance if fed at 2kg per day. It should be fed in hay racks to prevent waste. The rack should be large enough to allow the sheep to feed without overcrowding.

Ewes in the later stages of pregnancy will also require concentrates. Concentrates are highly nutritious food e.g. grain and various by-products from the edible oil seed crushing mills. These can be purchased in cube or pellet form to a guaranteed protein content from the local feed merchant. Introduce six weeks before lambing at the rate of 100g per day, increasing to 750g per day at lambing and afterwards. The concentrate is best provided as a compound mix of 17 per cent protein. It is important that the

Feeding hay during the winter months

Adequate feeding space

compound is one specially formulated for sheep, as those prepared for other classes of livestock can contain minerals in proportions injurious to sheep. Similarly, sheep that are being fattened for meat or being prepared for show or sale, may require supplementary food.

Besides good quality hay and the concentrates, it may be necessary to provide more bulk in their diet. All kinds of root and cabbage crops are suitable, although one should watch out for the possibility of choking. NEVER feed green potatoes.

Sugar beet pulp is a very useful food. When fed dry the analysis compares with rolled oats. When soaked overnight in water, the extra bulk is beneficial in giving extra substance to animals being fed for show. If mangolds are available, they should be fed to ewes only after lambing. Any secondary shoots which have grown during storage should be trimmed off before feeding to wethers or rams, as they will cause urinary problems.

Supplementary food can be expensive, and it is important that the sheep obtain the maximum benefit from the food offered. All the feed troughs and water containers should be kept clean and free from contamination from faeces, mud, etc. and there should be sufficient trough space to allow all the sheep to feed in comfort. Hay should be fed from racks or hay boxes and kept dry at all times. If racks are used, try to prevent the sheep from standing directly underneath as the hay seeds and chaff will contaminate the fleeces thus reducing the value of the shorn wool, and they may also be responsible for eye problems.

If you offer young lambs supplementary feed in the form of pellets specially formulated for them, this can be done by fencing off an area to which only the lambs have access. Vertical posts which are spaced so that only the lambs can pass between them, together with a horizontal rail 450mm from the ground, will effectively exclude adult sheep.

Provide access to a balanced mineral supplement in a waterproof container at all times. Check with your vet whether there are any particular mineral deficiencies in your area, and if so how to overcome them.

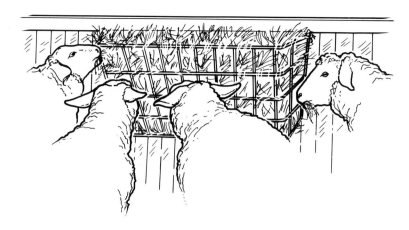

DIFFERENT TYPES OF HAY RACKS

ROUTINE TASKS

In addition to the general requirements listed above, there are specific tasks which periodically require attention. See page 105 for the equipment to use in performing these tasks.

Foot Trimming

Foot trimming becomes necessary when the horny outer surface of the foot grows beyond the soft, rubbery sole of the foot. Sheep that are housed and those kept on damp, heavy land will require more frequent attention than those kept on drier and more abrasive soils. Feet that are not trimmed regularly can become compacted with soil and other debris, eventually causing lameness. To trim the foot, first immobilise the sheep and using straight-bladed secateurs or a sharp knife, carefully pare away the overgrown horn down to the level of the sole of the foot. Sometimes the pointed front end of the foot will have grown forwards and upwards. Be particularly careful when paring in this area as severe bleeding will occur if it is cut back too far.

TRIMMING OVERGROWN FEET WITH FOOT TRIMMING SHEARS

Before (above) and after (below) trimming

Dosing

In order to keep the sheep free from internal parasites, they will need to be dosed periodically with anthelmintics. These can be in the form of liquids, paste or capsules. Some are broad spectrum, which can deal with a number of different kinds of parasites, while others are more specific. It is important that the manufacturer's instructions for dosing are followed correctly, and as dosage depends on weight it will be necessary to weigh representative individuals to ensure that all the sheep receive the recommended amount of the anthelmintic. Give each sheep a temporary crayon mark as it is dosed to avoid giving a double dose to some and missing others completely!

Sheep can be dosed in the standing position. Hold the head in a horizontal position by placing a hand under the lower jaw and insert the instrument containing the anthelmintic into the corner of the mouth and onto the back of the tongue before releasing the dose. Various types of dosing instrument are available. These include:

- A trigger-operated gun which automatically delivers a pre-determined measured dose from a reservoir carried on the operator's back.
- A paste gun which delivers a measured dose.
- A narrow-necked bottle which will require filling with a measured dose between each sheep.
- A large plastic syringe (minus the hypodermic needle!) which is more suitable for young lambs.

If the sheep starts to cough violently, release the head at once to avoid choking.

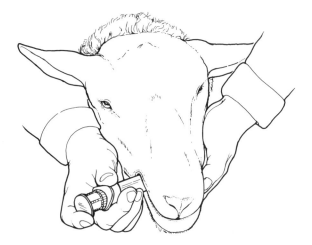

DOSING WITH A PASTE GUN

AND

WITH A TRIGGER-OPERATED GUN

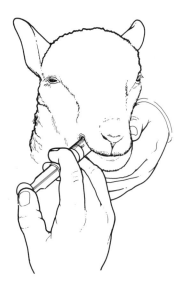

DOSING A LAMB WITH A SYRINGE

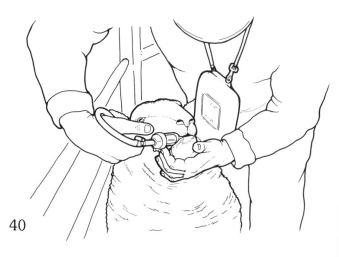

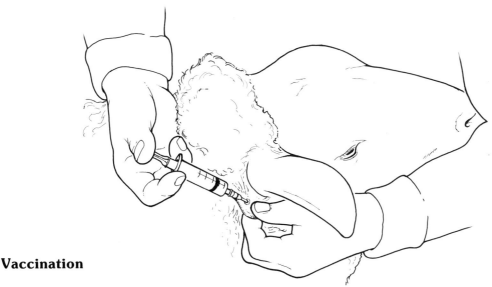

Vaccination

It is possible to control a large number of diseases affecting sheep by preventative vaccines, and if these are given at the correct times and in the correct quantities, the immunity to the diseases can be passed from the ewe to the lamb via the colostrum, giving protection to the lamb for up to ten weeks. Depending on the type, the vaccine can be administered in one of three different ways.

1. For a **subcutaneous injection**, lift a loose fold of skin with a thumb and forefinger, insert a hypodermic needle into the fold and release the vaccine by depressing the syringe. The preferred site for injections of small quantities of vaccine is high up on the neck, behind the ears. Quite often a small abscess or nodule will form at the injection site, and this can affect the subsequent value of the skin if present anywhere other than near the edge.

2. **Intramuscular injections** should be given at a site where there is good muscle cover, e.g. into the hind leg. The needle is pushed straight into the muscle and the syringe contents released slowly.

3. A **live vaccine** is applied to an area of bare skin between the front leg and the chest for protection against orf. The site should be clean but not disinfected otherwise the vaccine might be affected, and after application the site is lightly scratched so that the skin is broken to allow penetration of the vaccine. Gloves should be worn when using this vaccine as orf can also affect humans.

A SUBCUTANEOUS INJECTION (ABOVE) AND AN INTRAMUSCULAR INJECTION (BELOW)

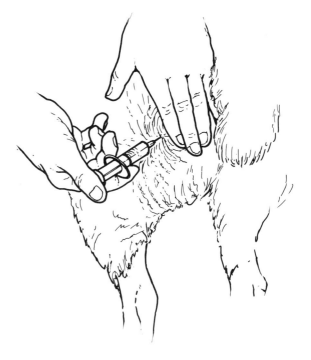

41

Dagging and Crutching

Both of these operations involve removing wool from the tail area of the sheep. In the spring when grass is growing quickly, the faeces are more liquid and can contaminate the wool. A build-up on the sheep can attract the blowfly to lay its eggs from which maggots will develop. Removing the wool helps to prevent this from happening, and it also avoids contamination of the fleece at shearing time.

A pair of sharp, short-bladed (100mm to 150mm) hand shears is used for dagging. The sheep is secured in a standing position by passing its head through a loop of rope attached to the side of the pen. Place a knee in front of the hip of the sheep, leaving both hands free to complete the operation.

Crutching involves taking off the wool from the area around the genitalia of the ewes prior to mating, but probably the main benefit will be at lambing time when the shorter wool makes visual inspection easier, particularly with longwoolled sheep. If wool is removed in front of the udder and from the front of the hind leg, it will give the newborn lamb easier access to the teats. To do this it will be necessary to turn the ewe as when trimming feet unless you possess a sheep cradle.

Dagging or crutching using hand shears

SECURING A SHEEP

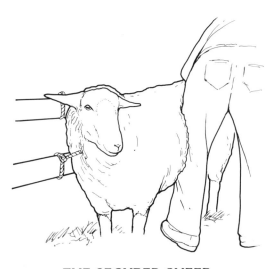

THE SECURED SHEEP

1

2

REMOVING THE WOOL
FROM THE TAIL END OF THE SHEEP
USING HAND SHEARS

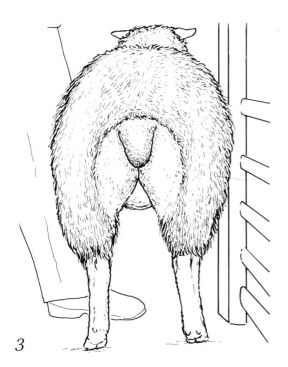

3

Shearing

In May or June all sheep that are more than
one year old will need to be shorn. This is a
skilled specialist operation and best left to
shearing contractors. It might be possible to
take the sheep to a neighbouring sheep
farmer when his flock is being shorn, but the
arrangements should be made well in
advance as the shearers are usually very busy
people.

Sheep are much less likely to suffer from
stress at shearing if they are penned and
kept away from food for at least twelve
hours before they are to be shorn. They sit
more easily on the shearing board, and the
board stays clean and free from faeces. The
wool *must* be dry before the sheep are
penned under cover. Shorn fleeces which are
wet, and contaminated with faeces, soil or
vegetable matter will attract a penalty from
the British Wool Marketing Board when the
wool is sold, as will fleeces tied with cord,
polypropylene or plastic twine.

If the flock consists of four or more

out on a flat clean surface and any part of the fleece contaminated with faeces, mud, etc. removed. The loose belly wool, if clean, is placed on the centre of the fleece and the sides are folded to the middle before rolling from the tail end towards the neck. Make a small hole in the fleece and tuck the neck wool into this, and place the whole fleece in the sack provided. (See illustration on page 99.)

Any skin cuts resulting from shearing will need treatment to assist healing. Keep a watch on any cuts that are being plagued by flies and treat again if necessary. The wound may be in a position where the sheep can strike it with its hind feet in an attempt to keep the flies away. This can soon result in an unpleasant open sore and it may become necessary to use lightweight sacking shaped like a horse rug to cover the body of the sheep. It should be inspected daily and dressed with an insecticidal cream. In extreme cases, putting the sheep into a darkened shed in the daytime will give it some respite from the unwelcome attention of flies.

Routine Inspections

All through the year a routine inspection of the flock is necessary, the frequency of the inspections varying with the time of year and the breed and condition of the sheep. Ideally it should never be less than twice a day, and more in the period preceding lambing. At this time the ewes are in good condition and heavy with lambs, and they can get on to their back when rolling and be unable to regain the standing position. This is common in early spring when showery weather alternating with drying sunshine and winds makes the sheep itchy. They roll onto their backs to relieve the itch and then get stuck in that position. If left like this, the continuing fermentation of food in the stomach and the pressure of the unborn lambs combine to restrict the breathing severely, eventually leading to suffocation and death.

ewes, some months before shearing it is necessary to register the flock with the British Wool Marketing Board, Registration Department, Oak Mills, Clayton, Bradford, West Yorkshire BD14 6JD. They will undertake to purchase the fleeces and will provide sacks to pack them in. Their agents will arrange collection after being notified that shearing has been completed. The British Wool Marketing Board requires that the fleeces are rolled in the correct manner. With shortwools the skin side of the fleece is on the outside, and with longwools the skin side is on the inside. The shorn fleece is laid

In the period from early May until early November the sheep maggot fly is active and a careful watch is necessary for any sheep that are 'struck'. These will require prompt treatment so that any injury to the sheep caused by the maggots burrowing into the skin is minimised. A few weeks' protection from the maggot fly is obtained when ewes are newly shorn and also when the lambs have been sprayed with an insecticidal dip.

Spend time watching the sheep without disturbing them so that any abnormal behaviour can be quickly recognised and if necessary acted upon.

Teeth

There are no incisor teeth in the sheep's upper jaw! Eight small incisor teeth appear soon after birth in the lower jaw. At about fifteen months of age the centre two are shed, and replaced by two larger, permanent incisors. At yearly intervals a tooth on either side of these will be shed and replaced by permanent ones, and the sheep is known consecutively as a two-tooth, four-tooth or six-tooth, according to the number of permanent incisor teeth present. At just over four years of age an adult sheep should have eight permanent incisor teeth and is then known as full-mouthed.

After four years the teeth continue to grow and will eventually be shed. The sheep is then said to be broken-mouthed. This may lead to feeding problems and unless special provision can be made any such ewes should be culled before mating.

Occasionally sheep are born with a lower jaw shorter or longer than normal. This can lead to feeding difficulties as the teeth will not meet the upper jaw pad correctly, and is an undesirable characteristic. Sheep with this condition are said to be 'undershot' or 'overshot' and should not be used for breeding. Careful examination of the teeth of any sheep which you consider buying is necessary.

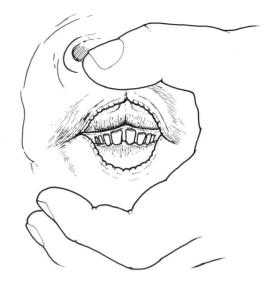

THE INCISOR TEETH OF A YOUNG LAMB

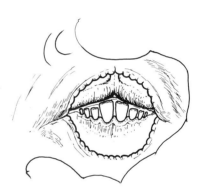

OF A 15 MONTH OLD SHEEP

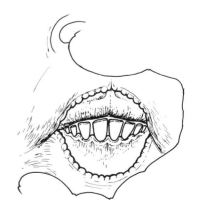

OF A 3 YEAR OLD SHEEP

45

A broken mouthed sheep

BODY CONDITION SCORING

Condition scoring is an attempt to standardise how much fat and muscle a sheep is carrying by using the fingers to assess the degree to which the vertical and horizontal processes of the backbone are covered. The point where the assessment is made is in the region of the loin, behind the last rib and over the kidneys. It is scored on a scale of 0 to 5, and half points are also used.

Although the flock may be all of the same breed or cross, there will be considerable variation between individuals at any particular time. Between different breeds the variation will be even greater, and it will prove to be almost impossible to get all the animals into the same condition at the same time even if they are kept under the same conditions!

Any problems with the molars, or back teeth, in either upper or lower jaws can usually be felt as hard lumps through the skin along the line of the jaw. Your vet will advise of any treatment necessary.

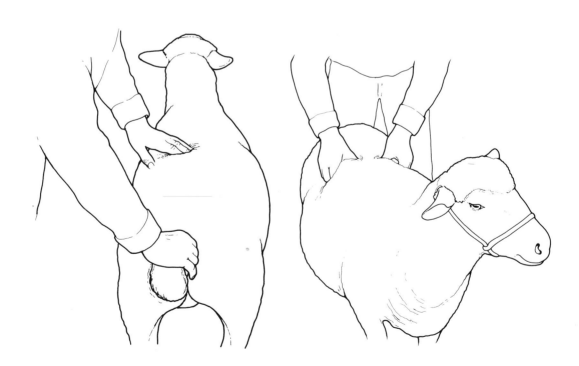

ASSESSING THE BODY CONDITION AROUND THE DOCK AND OVER THE LOIN

THE POSITION OF THE HANDS ACROSS THE LOIN

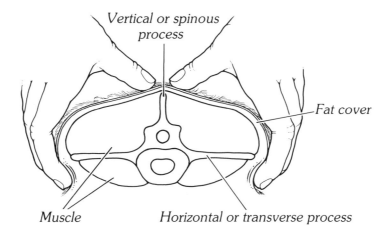

Vertical or spinous process

Fat cover

Muscle

Horizontal or transverse process

Score 0 This indicates extreme emaciation with no fat or muscle detectable.

Score 1 The backbone feels very sharp in both the vertical and transverse processes, and muscle tissue in the angle between is very shallow with no fat cover.

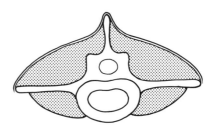

CONDITION SCORE 1 — LEAN

Score 2 The muscle tissue is of sufficient depth to give the processes a smooth and rounded feeling, but there is little fat cover.

Score 3 Individual spinous processes can be felt only by pressure and are smooth and well covered. The loin muscle is full with a moderate fat cover.

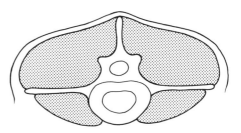

CONDITION SCORE 3 — GOOD

Score 4 The loin is full with a thick covering of fat and the spinous processes detectable only as a firm line. The ends of the transverse processes cannot be felt.

Score 5 The animal is very fat. There is a shallow groove in the fat cover over the vertical processes and the transverse processes are undetectable.

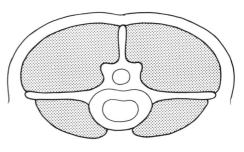

CONDITION SCORE 5 — FAT

CHAPTER 6

Caring for the Breeding Flock

CARE OF THE BREEDING EWE

A typical year for the breeding ewe begins a few weeks before mating takes place. The planned size and composition of the flock will have been completed, and the ewes brought on to a rising plane of nutrition, aiming for a body condition score of 3.5 at mating time. The prolificacy of the flock will be increased if the ewes are 'flushed' about three weeks before mating. This involves putting them onto better grazing if it is available, or feeding 200g of concentrates to each ewe each day until mating is completed. Flushing will result in more eggs being shed by the ovaries and more lambs being born. Any parasitic or foot problems will have been corrected, and any young sheep will have been given the appropriate vaccine in order to protect them from the clostridial group of diseases and from pasteurella pneumonia.

When the ram has joined the flock, and mating has begun, all unnecessary stress for the ewes should be avoided. It is best if the ram is fitted with a harness with a marker crayon which puts a colourful mark onto the rear of each ewe as she is mated, as this will

Ewes in good condition

enable you to keep a record of the mating date and work out the expected lambing date for each ewe. The gestation period is about 147 days.

When mating has been completed, remove the ram to a separate enclosure if possible. The condition of the ewes can now be allowed to drop back to a score of 3, although they should be given hay to appetite when the quality of the grazing deteriorates. If housing is available and the weather is bad, consider housing the ewes but try to ensure that the fleeces are dry before they move into the house for any period of time. Wet sheep will take a long time to dry when under cover.

Six weeks before lambing commences, begin concentrate feeding in addition to hay at the rate of 100g per sheep each day. Increase this amount weekly so that each ewe is receiving 750g daily when lambing commences. A booster injection of the vaccine against clostridial diseases should be given four weeks before they are due to lamb.

When each ewe has lambed place her in an individual pen made with hurdles and measuring 1.8m square. Any size smaller than this will expose the newborn lamb to the risk of being lain on when the ewe lies down, especially with multiple births. She should stay in the pen with her lambs for a few days until it is apparent that all is well with both ewe and lambs. The afterbirth will have come away shortly after lambing, and it should be collected and disposed of away from predators. The udder of the ewe will need to be inspected to ensure there is an adequate supply of milk for the lambs, and the lambs castrated and docked using rubber rings. When turning the ewes out of their individual pens with their lambs to join others that have already lambed, it is advisable to give each family an identification mark which will allow them to be recognised easily if for any reason they have to be segregated, or if a lamb loses sight of its mother and has to be reunited with her.

This is also a good opportunity to inspect each ewe's feet and trim if necessary, and also to give a dose of an anthelmintic to make sure she goes out worm-free. This dosing should continue at monthly intervals until the lambs are weaned.

During the lactation, and especially after weaning, keep a careful watch on the ewes for any sign of mastitis in overstocked udders, but do not be tempted to strip milk from ewes with healthy udders after weaning. This will only encourage the udder to produce more milk and will hold up the natural process of reabsorption.

If the ewe has successfully reared two or more lambs to weaning, she will very probably have lost condition and have a body condition score of around 2.5. Provided that she is healthy and that there is sufficient time to bring her up to a score of 3.5 before mating again, this is no cause for concern.

Before making up the breeding flock for the following year, a careful inspection of all the existing flock members is required to make sure that no unproductive passengers are being carried. Start with a look at the incisor teeth. Are the teeth present compatible with the age of the sheep? Do any sheep over four years of age have any incisors missing? If they have, and in all other respects they are sound and healthy, will facilities be available for them to have sufficient supplementary food in order to maintain their body condition, especially in the later stages of pregnancy? Feel the outside of the jaws along the site of the molars (back teeth) for any hard lumpy tissue which will indicate infected and possibly decaying teeth which will require veterinary attention. Next handle the udder, feeling for any lumps of hard tissue which will indicate chronic mastitis. A few weeks after weaning the udder should have returned to a soft, pliable consistency.

Any ewes with one or more of these defects or any adults that have not bred lambs when given the opportunity to do so for two consecutive years, should be given a separate identifying mark for culling. This also applies to any suffering from recurring chronic illnesses and those with chronic foot problems. Any ewes which have required a caesarean section operation to deliver their lambs will have only a fifty per cent chance of breeding successfully in future years, and their retention as flock members should be carefully considered.

Following this inspection the sheep marked as culls can be marketed for meat when the price offered is favourable. Any surplus ewes suitable for breeding can be advertised for sale or taken to sheep auction sales which take place from mid-August until November.

Final decisions on the size and composition of the flock can then be made a few weeks before mating is due to begin. Any additions or replacements will have been made well in advance.

SELECTION AND CARE OF THE RAM

The breed of ram selected to be the sire of the flock is one of personal preference, but certain basic requirements in the conformation are common to all. Look for:
1. Good feet and straight legs with an easy walking action
2. Correct mouth with neither undershot nor overshot jaw
3. Bold eye and alert expression
4. Large, even-sized testicles, firm to the touch
5. Long, straight back and good spring of rib

Pay particular attention to the choice of ram, looking for one at the breed standard for size and weight. Always buy the best you can afford. If the ewes of the flock have a particular fault which needs correcting,

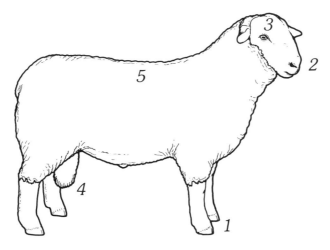

POINTS TO LOOK FOR IN A GOOD RAM

beware of using a ram which is particularly good on that point while weak on others. Remember that the ram is responsible for half of the genetic make-up of his progeny.

Try to arrange for the ram to be kept separate from the rest of the flock until he is required for mating, but if possible give him company by providing him with a wether, or a ewe that you know has been successfully mated. A ram kept by himself without the company of other sheep can become aggressive, although this aggression can be minimised by fitting a face mask which restricts forward vision. Avoid making a pet of the ram by giving him titbits, as this too will encourage aggression.

Use a harness when the ram is with the ewes. The mark of a coloured crayon fitted to the harness will identify each ewe when mating takes place, and will let you know when to expect the lambs to be born. The harness should be fitted to the ram without the crayon a few days before mating is due to start to allow it to settle into the fleece, and if necessary it can be tightened to give a good firm fit when the crayon is added.

Check periodically that the harness and crayon-holder are not chafing.

The crayon should be changed after seventeen days to another colour as a check that mating has been successful, as this is the interval when the ewe's heat cycle recurs. If more than one or two ewes are marked by

Ram mask

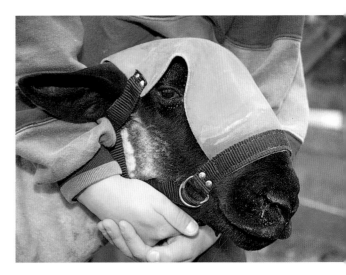

Ram wearing a harness with his ewes

the second colour, having already been marked by the first one, it may be that the fertility of the ram is questionable, but wait a few days longer to see whether more ewes return for a second service before considering whether it is necessary to use another ram. Another change of colour should be made after a further seventeen days, by which time all the flock should have been mated.

The ram should always be kept in good condition, fit but not overfat (condition score 3.5 to 4), and in good health. Pay particular attention to the feet, inspect regularly and trim as necessary. A lame ram will be reluctant to work and this can mean a reduced lamb crop.

It is difficult to say with accuracy how many ewes a ram can successfully serve in a season, but an optimum number is around forty-five. Mature rams from very active breeds like the Wiltshire Horn are capable of serving many more. Avoid overworking ram lambs.

If the size of the flock does not justify the cost of purchasing a ram, it may be possible to hire one just for the mating season, or to have an arrangement with another flock owner where the ewes can be taken. Bringing other sheep in or sending sheep away for a short time can bring problems, e.g. footrot and parasitic infection. Have a careful look before exposing the sheep to the possibility of infection from outside sources.

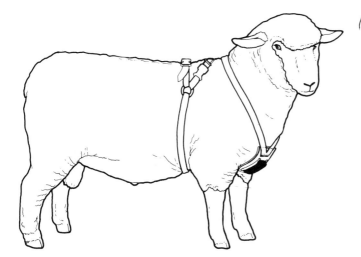

THE RAM HARNESS

CHAPTER 7 **Lambing**

Lambing time can be the most rewarding, demanding and sometimes the most stressful time of the year for both sheep and shepherd. It is the culmination of all the efforts to ensure that the health, nutrition and general wellbeing of the sheep result in a successful increase in the size of the flock.

By far the largest proportion of ewes lamb successfully unaided but a small number do need assistance. The way in which this assistance is given can vary a great deal, but the methods described in this book have been developed and proven over many years by us to give acceptable results.

It will not be possible to save every lamb that is born, as some may be still-born, but the golden rules are:

- Have your lambing kit to hand well before lambing commences, and also the telephone number of your vet.
- Practise as high a standard of hygiene as possible under the circumstances prevailing.
- Be patient. Do not rush to assist even in a normal birth, especially with ewes lambing for the first time. It is possible for the whole process to last for four hours and still result in a normal, natural birth.

The lambing kit: (1) lubricant; (2) baby feeding bottle and teats; (3) stock marker; (4) calcium borogluconate; (5) navel dressing; (6) pessaries; (7) torch (8) lambing ropes; (9) castrating pliers with rubber rings; (10) frozen colostrum; (11) stomach tube with syringe; (12) hypodermic syringe with needles; (13) vaginal support; (14) sheep halter; (15) lambing aid

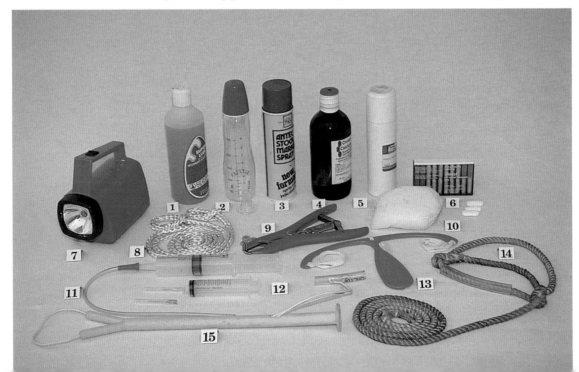

THE LAMBING KIT

You may require the following lambing aids:

● *A lubricant*   This is for use when it is necessary to assist a ewe by an internal examination and/or manipulation of the unborn lamb. This can be in the form of a mild antiseptic oil or cream, or soap and water. Soap flakes are particularly good.

● *A 1m length of thin cord*   By tying a loop at each end this can be used to secure an unborn lamb's feet or head and will make it easier to pull the lamb from the ewe should this become necessary. The cord should be stored in a weak antiseptic solution in an unbreakable container.

● *A lambing aid*   This is a Y-shaped plastic tube containing a retractable plastic covered cord. It can be used, with experience, to secure the unborn lamb's head and feet together inside the ewe and therefore enable the operator to withdraw the lamb from the ewe in a normal presentation position.

● *A navel dressing*   This dressing dries the lamb's navel very quickly and thus prevents harmful organisms gaining entry to the lamb, particularly the bacteria that cause joint-ill and liver abscess.

The dressing can be:
> Tincture of iodine
> Antibiotic aerosol
> A very weak solution of copper
> sulphate.

Ideally, the dressing should be kept in a wide-necked container so that the navel cord and surrounding area can be immersed by holding the lamb vertically by the front legs and pressing the container gently against the lamb.

● *A supply of colostrum*   It is absolutely essential that newborn lambs are given colostrum as their first feed. Colostrum is a sticky yellow substance usually present in the udder when the lamb is born. It is a rich source of energy and also a laxative;

provided the ewe has been given the appropriate vaccines it will contain antibodies which will give the lamb protection from specific diseases through the early stages of its life.

If the ewe has not produced any colostrum, or if there are more than two lambs, a supply can be obtained from a ewe whose lambs have died at birth or from one side of the udder of a ewe that has produced a single lamb.

If neither of these sources is available, colostrum from a newly kidded goat or a newly calved cow can be substituted, but unless the animals have been vaccinated beforehand it will be necessary to vaccinate the new-born lamb against clostridial diseases.

Colostrum can be frozen in small packs of 200ml for storage. It must be properly thawed and heated to 40°C before use.

A final alternative is to use a ewe colostrum replacer made up according to the manufacturer's instructions.

● *A stomach tube with syringe*   This is necessary when giving colostrum to weakly lambs.

● *Baby feeding bottles and teats*
Sometimes the ewe's supply of milk in the first few days is insufficient for her lambs. If so, after the initial feed of colostrum, the lambs will need supplementary feeding with milk from other ewes or with a good-quality ewe milk replacer.

● *Rubber rings*   These are used together with the appropriate pliers for tailing and castrating.

● *Pessaries*   These are either antibiotic or antiseptic and are placed in the uterus of the ewe after her lamb has been delivered following internal manipulation. They can be obtained from your vet.

● *A hypodermic syringe and needles*
These are necessary for injecting the ewes and lambs.

- **A quantity of magnesium fortified calcium borogluconate** This is injected subcutaneously into ewes which collapse suddenly with hypomagnesaemia (magnesium deficiency) or hypocalcaemia (milk fever).

- **A wire or plastic support for vaginal prolapse** A vaginal prolapse usually occurs during the few weeks before lambing. Normally it first appears as a pinkish rose-shaped object, and may not be showing all the time. If the condition deteriorates and the prolapse gets bigger, then the ewe may have difficulty in urinating. This causes strain, and in the extreme condition the wall of the vagina may break and consequently the intestines escape through the hole. When it reaches this stage there is no cure and the vet should be called to destroy the ewe. The wire support, which has loops to secure it to the fleece of the ewe, should be inserted into the vagina when it is apparent that the prolapse is not going to go back on its own. It can be left in position until lambing is well advanced and the lamb is about to be born.

- **Stock markers** It is advisable to have several different colours available to mark lambs or ewes needing to be identified temporarily.

- **A halter** Occasionally a ewe will reject one lamb of her twins or triplets and a short spell tied with a halter will usually overcome this problem. The halter rope should be tied to a corner of the pen, high enough and short enough to prevent the ewe from turning round, but still allowing her to feed and drink easily. It is also a good idea to put a length of rubber tubing on the part of the halter which passes under the jaw. This will prevent the halter from becoming tight and therefore causing chafing.

- **A torch** with a spare battery

- **Disposable plastic gloves**

- **The vet's telephone number**

LAMBING: SIGNS AND OBSERVATIONS

The first indication that a ewe is about to commence lambing is that she behaves abnormally. Spend time watching normal behaviour in ewes so that any abnormal behaviour can be recognised.

The ewe will select a spot for her lambs to be born, and if she is driven away from it she will return at the earliest opportunity. She will become agitated, often pawing the ground and bleating softly. She will also lie down and get up again at frequent intervals. It is advisable to allow the ewe to remain at the site she has chosen, provided that it is sheltered, safe and dry. She is more likely to get on with the job of producing her lambs

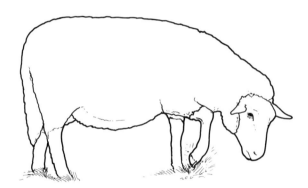

FINDING A SPOT TO LAMB

THE EWE STRAINS

Early signs

Selecting a spot

The ewe begins to strain

than if she is moved to another site or put in a small pen. Some ewes do not like an audience at close quarters when lambing and may give up attempting to lamb if anyone is near.

Labour

The ewe begins to strain very obviously as the birth commences and she lies down, often with her legs stretched out to one side. She often raises her head, closes her eyes and curls her top lip. The 'waterbag' is a transparent membrane containing the birth fluid. It may burst inside the ewe, but usually appears outside and bursts within a few minutes.

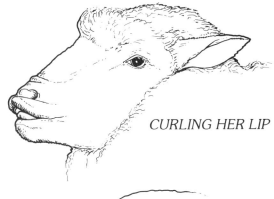

CURLING HER LIP

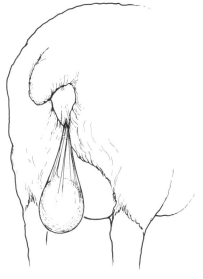

THE WATERBAG APPEARS

A normal presentation

With continued straining the lamb's nose and both front feet will appear. Often they will be enclosed in a transparent membrane. Slowly the body of the lamb will slide from the ewe. If the transparent membrane is still intact, prompt action is necessary to remove it; otherwise the lamb will be unable to breathe and will suffocate.

The nose and both front feet appear

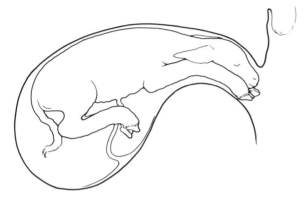

A NORMAL PRESENTATION

The lamb will kick and cough and then the ewe will stand and turn to lick it. This has the combined effect of drying the lamb, establishing a bond between lamb and ewe and stimulating the lamb's circulation. This is a time of rapid heat loss in the lamb, particularly in inclement weather, and careful watch should be kept to see that it does not suffer from hypothermia due to exposure. If the ewe is reluctant to start licking, place the lamb at her head and rub some of the birth fluid over her nostrils.

After the birth of the lamb, the ewe can be moved to a sheltered position by holding the lamb by front or back legs and drawing it slowly in the direction that you wish the ewe to travel. It is very important that the lamb stays at ground level where the ewe can see it and can continue licking and smelling it as

The lamb slides from the ewe

The ewe licks the lamb

she moves. If she loses sight of the lamb she will return to the spot she chose for the birth.

Before leaving the ewe with her lamb, gently draw the thumb and forefinger down her teats to break the seal at the end so that the lamb can easily obtain the colostrum. The afterbirth should come away from the ewe after the birth(s) have been completed. Some ewes will attempt to eat the afterbirth, but it should be cleared away as it can distract the ewe from caring for her lamb. It will also attract predators — dogs, foxes and carrion crows. If the afterbirth becomes covered with straw and is allowed to decompose for some time, the smell of decomposition can be most unpleasant. It should be properly disposed of by burying.

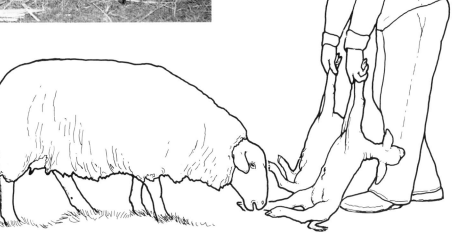

MOVING THE EWE AND NEWBORN LAMBS

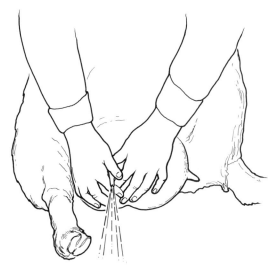

CHECKING THE UDDER

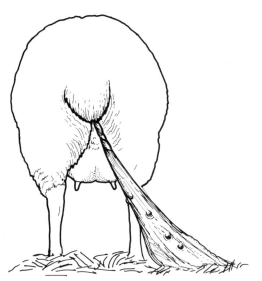

THE AFTERBIRTH OR CLEANSING COMING AWAY

Whether to assist the birth

Patience is a virtue! Do not rush to assist the ewe even in a normal birth. If, after an hour or so of labour commencing, nothing can be seen of the lamb and no progress is being made, then quietly catch the ewe. Insert one or two fingers into the vagina and try to identify any part of the lamb which you can touch. If an assistant is available to hold the ewe, this can be done while the ewe is still standing. Otherwise, it will be necessary to immobilise her by laying her down on her right side. This can be done by standing or kneeling at her left side and pulling her head round towards you while simultaneously reaching under her and grasping her right hind leg, then pulling it through in front of her udder towards you.

This is a move that can be practised at any time of the year, prior to the lambing season. Use a small sheep and progress to a larger one as you become proficient. The ewe will collapse onto her right side and your left hand and left leg can be used to keep her immobilised while the right hand makes a preliminary examination.

Remember that a normal birth can take up to four hours and if your examination shows that no part of the lamb can be felt and entry to the uterus is dry and restricted, then release the ewe and wait a while before making another examination. If after three hours no further progress is being made, **do be patient** and **do not force** an entry. Seek help from your vet.

Ringwomb

This is a condition where the cervix, or entrance to the uterus, fails to dilate. It can be recognised when a preliminary examination reveals, as the name implies, a tight inelastic ring at the entrance to the uterus. Do not force an entry but contact your vet.

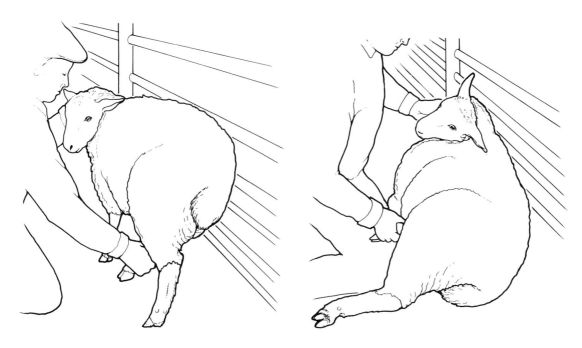

IMMOBILISING THE PREGNANT EWE

KEEPING THE EWE IMMOBILISED

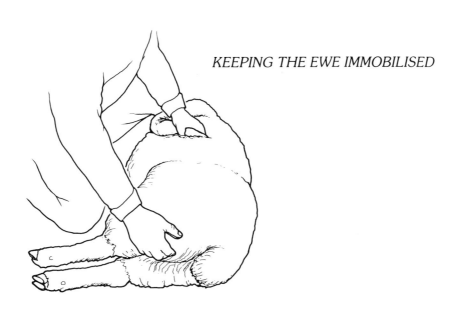

Malpresentations

Unfortunately a number of malpresentations can occur, and if the examination shows that the lamb is being presented in a position other than normal, then one of the procedures on the following pages is called for. *If you feel that you are unable to cope with these malpresentations, seek help from your vet before attempting to proceed beyond your initial examination.*

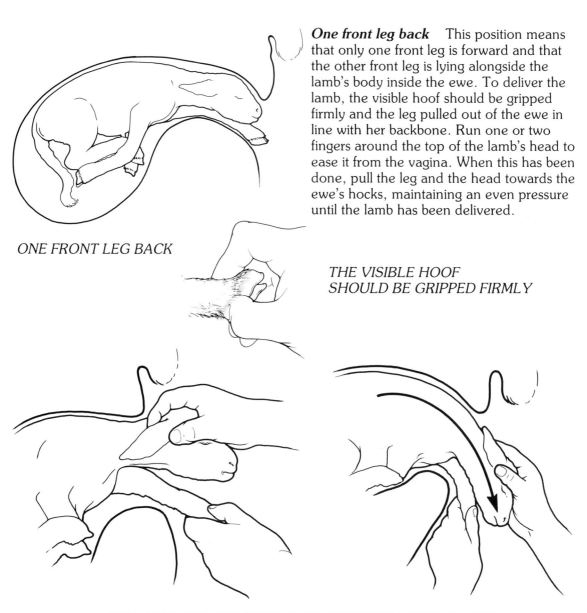

One front leg back   This position means that only one front leg is forward and that the other front leg is lying alongside the lamb's body inside the ewe. To deliver the lamb, the visible hoof should be gripped firmly and the leg pulled out of the ewe in line with her backbone. Run one or two fingers around the top of the lamb's head to ease it from the vagina. When this has been done, pull the leg and the head towards the ewe's hocks, maintaining an even pressure until the lamb has been delivered.

ONE FRONT LEG BACK

*THE VISIBLE HOOF
SHOULD BE GRIPPED FIRMLY*

PULL THE LEG AND HEAD TOWARDS THE EWE'S HOCKS

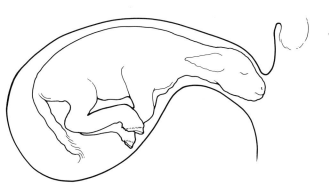

BOTH FRONT LEGS BACK

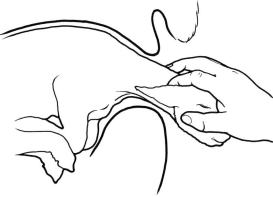

EASE LAMB'S HEAD OUTSIDE THE EWE

Both front legs back   In this case ease the lamb's head right outside the ewe and slide the right hand down the left side of the lamb's neck, feeling for its left shoulder and left front leg. Crook a finger behind the leg and pull it up in line with the lamb's neck. Deliver the lamb by pulling the foot and head downwards towards the ewe's hocks. It is not necessary to re-insert your hand into the ewe to bring up the other front leg.

   If the ewe has been unattended and in labour for some time, she may have succeeded in getting the lamb's head out of the vagina. When this happens its head and tongue may appear very swollen but the swelling will quickly subside once the lamb has been delivered.

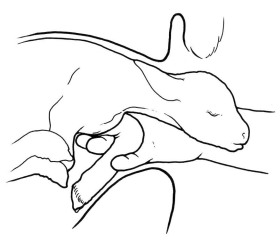

LOCATE A FRONT LEG AND PULL TO THE OUTSIDE. DELIVER AS WITH ONE FRONT LEG BACK

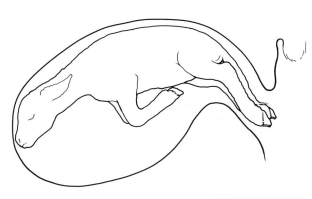

HIND LEGS FIRST

Hind legs first   If, on examination, only two feet can be felt and no head, and the points of the hooves are on the underside, then feel up the legs for the lamb's hocks. If they are present grip them firmly and quickly deliver the lamb by pulling it out in line with the ewe's backbone. It might be necessary to put a cord around the feet of the lamb to enable you to get a better grip on them, but it is important that the lamb is delivered quickly. If not, there is a possibility that it will attempt to breathe when the umbilical cord breaks while the head is still inside the ewe.

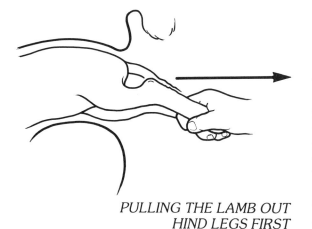

*PULLING THE LAMB OUT
HIND LEGS FIRST*

It will then take the birth fluid into its lungs and may suffocate. If you suspect that this has happened, hold the lamb by its hind legs and swing it violently backwards and forwards in an attempt to expel the fluid and mucus from its lungs. Clear away any mucus from the lamb's mouth and check whether its heart is beating by gently pressing on its rib cage immediately behind its shoulder. If the heartbeat is present but the lamb is not breathing, mouth-to-mouth resuscitation may help. If the need arises, don't be put off by the appearance of the lamb — there is virtually no taste to the birth fluid!

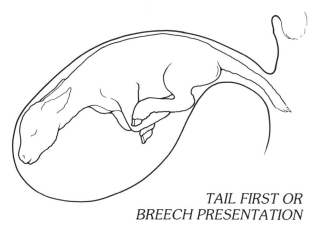

*TAIL FIRST OR
BREECH PRESENTATION*

Tail first or breech presentation   An indication that the ewe is attempting a breech birth will be the appearance of the lamb's tail. Gently push the hindquarters of the lamb back inside the ewe and reach for its hind legs. Pull both hind legs up and outside the ewe, and then quickly pull the lamb out in line with the ewe's backbone. If necessary, swing it and proceed as with hind legs presentation to expel fluid and mucus from the lungs.

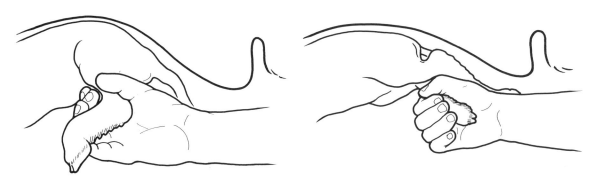

DEALING WITH A BREECH PRESENTATION

Head back, both front feet forward

This is the most difficult presentation to deal with. Firstly, identify positively that the feet are front feet by noting that the points of the hooves are uppermost and feeling for a knee. Do not under any circumstances pull on these legs hoping that a head will follow, as if the head is turned back, pulling the legs will force the head even further back.

Instead put the lamb's feet in the palm of your hand and, after attaching a thin, disinfected rope to each, push them back inside the ewe. As soon as it is possible to do so, push them down so the legs are under the body of the lamb. Locate the curved body of the lamb and pull the head forward by putting the thumb and forefinger in the eye sockets. Bring the head right outside the ewe, using the rope to bring up its legs at the same time. Deliver the lamb by pulling its head and legs towards the ewe's hocks.

This is an occasion when the lambing aid can come into its own. It can be used to snare the lamb's head and legs after locating them inside the ewe, and also for pulling them through the pelvis to the outside.

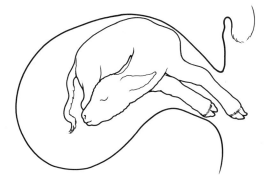

HEAD BACK AND BOTH FRONT FEET FORWARD

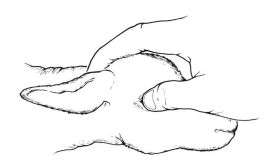

BRINGING THE HEAD FORWARD

Twin lambs presented together

This can be a puzzling situation for beginners. If two heads and some or all of the front legs are presented together, it will be necessary to push one head and all the legs back inside the ewe. Bring one head to the outside before going back for one front leg, making sure that it belongs to the same lamb as the head! Deliver the lamb by pulling the head and feet downwards towards the ewe's hocks.

In all other cases of mixed-up presentations, try to identify positively any part of a lamb that you can touch. If a head is presented, then work with this one. If at any time it is difficult to move the lamb inside the ewe, it sometimes helps to roll the ewe over onto her left side and then back again onto the right.

Guarding against infection

In all cases of assisted birth an antiseptic or antibiotic pessary should be inserted into the ewe when all the lambs have been delivered. Additionally, an intra-muscular injection of a long-acting antibiotic should be given to prevent infection. This is obtainable from your vet.

Dead lambs

You may find that you are having to assist in the birth of a dead lamb. This can be a difficult and unpleasant task, especially if the lamb is beginning to decompose. The dead lamb will feel dry to the touch and the natural lubrication of the birth will be absent. Use copious amounts of lubrication, soap and water for preference. If you feel you are unable to cope with this, consult your vet. He will also advise you on the use of long-acting antibiotics and pessaries.

CARE OF THE LAMB

The first few days of a lamb's life are critical to its future development, and a careful watch should be kept on all newborn lambs so that any potential problems can be recognised early.

Following a successful birth the ewe normally begins to lick the lamb clean and dry. The lamb gets to its feet after a few minutes, makes it way to the ewe's udder and begins to suck vigorously. The ewe and lamb should be placed in a small pen (1.8m × 1.8m) on a clean bed of straw. This will prevent the lamb from straying, especially in the case of a multiple birth where the ewe might have to care for several lambs. The lamb's umbilical cord should be dressed (see the lambing kit list on page 55) to guard against infection.

If the lamb has been born outside in inclement weather, it may be suffering from hypothermia, or chilling, and finding it difficult or impossible to get to its feet. Help to stimulate circulation by rubbing the lamb briskly with a dry towel, and introduce it to the ewe's milk by opening its mouth and inserting a teat, gently squeezing it to give the lamb a taste. This task is made easier by casting the ewe onto her side, and then by leaning over her and placing an ankle over her neck, which will leave both hands free to

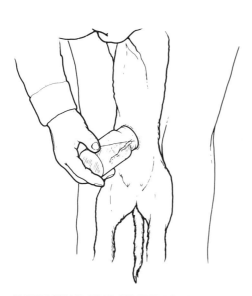

DRESSING THE UMBILICAL CORD

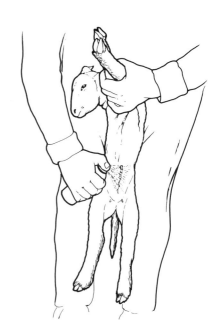

handle lamb and teat. An elbow gently placed into the flank of the uppermost leg of the ewe will prevent a kick in the face! If the lamb is reluctant to suck, it can sometimes be stimulated by scratching its tail head.

If it is still cold, wet and lethargic but able to hold up its head unaided, it can be fed using a stomach tube. To do this, sit comfortably with the lamb on the lap. Introduce the tube minus the syringe into the lamb's mouth and gently slide it into the stomach, making sure that it is in the oesophagus and not the windpipe. Place two fingers across the underside of the neck and if the tube can be felt, it is going down the oesophagus. If not, it is in the windpipe and should be withdrawn immediately. If the lamb shows no sign of discomfort and chews the tube when it is in position, this indicates that it is in the stomach and all is well.

In a large lamb all but two inches of the stomach tube can be introduced. The syringe should be filled with colostrum taken from the ewe and placed on the end of the tube. Depress the plunger of the syringe very slowly and repeat the process until the lamb has received between 100 and 200ml, repeating this four hours later if necessary. As soon as the lamb begins to suck unaided, place it at the ewe's teat.

Ensure that the lamb gets an adequate supply of at least one litre of colostrum during the first twenty-four hours of its life. If the ewe has no colostrum, use a baby's feeding bottle to give it from another source (see page 55).

If, after twenty-four hours, the ewe has no milk or insufficient milk, feed the lamb with milk from another ewe or use a good quality ewe milk replacer.

Apply a strict hygiene routine to the stomach tube, syringe, bottle and teat used to feed lambs, and thoroughly clean and sterilise between each feed.

If it becomes necessary to warm a lamb by an artificial heat source, several methods can be used. An infra-red lamp is useful, but

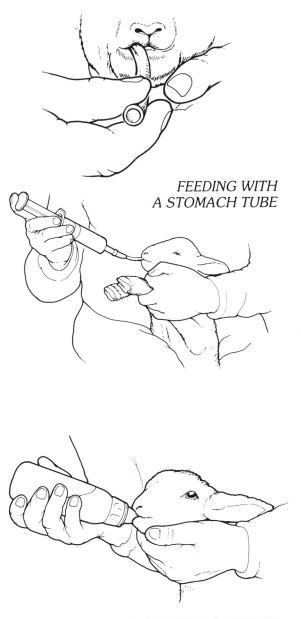

FEEDING WITH A STOMACH TUBE

FEEDING WITH A BOTTLE

take care to see that it is far enough away from the lamb (a minimum of 1.2m) to prevent scorching or burning. A drawback to this method is that when dry the lamb takes on a different smell and because the mother of a newborn lamb identifies it by smell alone she sometimes refuses to accept that it belongs to her. A way around this problem is to suspend the lamp above the lamb in a small fenced off corner of the pen where the ewe can still see and smell the lamb but where she cannot be burned.

A domestic hot air blower can be used to blow warm air onto the lamb lying in a box or confined space, as can the oven of a domestic cooker at a low temperature, but remember to leave the oven door open!

If no artificial heat source is available, a hay box can be used to revive a lamb suffering from exposure. Use soft hay to make a deep nest inside a container such as a large dustbin or forty-five gallon drum. Place the lamb in the nest and cover with clean hessian sacks or more hay before putting a lid on the container. An hour or so later the lamb will usually be standing on the sacks or hay and calling for food.

Any lamb which is cold and lethargic when several hours old and already dry should not be subjected to artificial heat. Give a feed of colostrum from a baby's feeding bottle or through a stomach tube if the lamb is reluctant to suck.

Castrating

When it is apparent that all is well — the lambs are receiving an adequate supply of milk and the ewe is mothering properly —

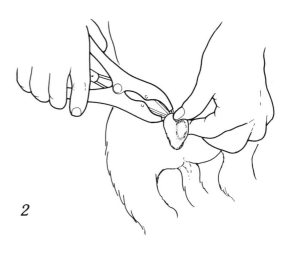

2

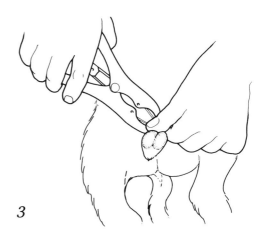

3

STEPS IN CASTRATING

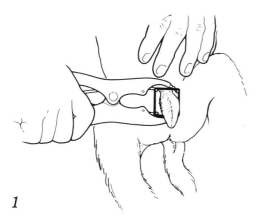

1

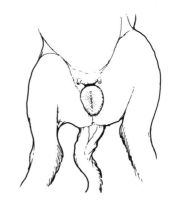

4

the ram lambs can be castrated. The simplest way for the beginner to do this operation is to use rubber rings and the appropriate pliers, called elastrators. By law this method is permissible only during the first week of a lamb's life, and by persons over seventeen years of age. First check for abnormalities. The lamb may be suffering from a scrotal hernia, when a part of the intestine is present in the scrotum with the testicles, and can be detected as a soft, spongy mass. Lambs with this condition will die if castrated. Sometimes either one or both of the testicles have not descended into the scrotum. If only one has descended, do not castrate. The tail should be left long for identification while waiting for the testicles to come down normally.

Kneel or sit with the lamb lying on its back between your legs and facing away from you. Place a rubber ring on the elastrator, stretch the ring open and draw the scrotum through it. Close the ring on the scrotum and then open it just sufficiently to allow a testicle to be eased through it into the scrotum, using the free hand (see illustrations on facing page). Close it again and repeat the procedure for the second testicle. Make sure that the rubber ring is well below the rudimentary teats and that both testicles are secured before removing the elastrator from the ring.

At sixteen to twenty weeks of age an uncastrated ram lamb becomes sexually active and can be a nuisance if left with the flock. Either segregate it or send it to market.

If it becomes necessary for any reason to remove the ring from the scrotum, insert a small piece of wood under the ring and cut through the ring onto it. This will avoid cutting the skin.

Docking

To dock the lamb place a ring on the elastrator, open the ring and pass the tail through it. Before closing the ring on the tail, ensure that sufficient tail is left above the ring

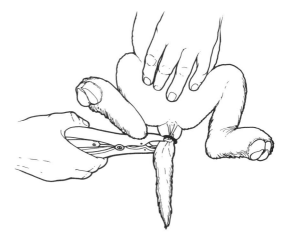

DOCKING

to cover the anus of ram lambs and the vulva of ewe lambs. (This is required by law.)

After these operations both ram and ewe lambs suffer some discomfort and will roll around on the ground for a short time. Because of this it is advisable to do the job either on a clean straw bed or on a clean grass field. The discomfort soon passes and the lambs are soon behaving normally again.

A good tail length

FOSTERING LAMBS

Occasionally it may be necessary to foster an orphan lamb onto a ewe that has lost her lamb, or to foster a lamb from a multiple birth onto a ewe that has produced only a single lamb. There are several ways of doing this.

Skinning

In the case of a dead lamb remove the skin, trying if possible to leave the holes for the legs intact. Place the skin with the tail still attached onto the lamb to be fostered, putting the legs through the holes left in the skin so that it covers all the body except the head and ears. If necessary, use string to join the skin to make a good fit. As a safeguard, put a halter on the ewe and tie her to the corner of a small pen. Tie the rope at a height that will allow the ewe to lie down in comfort but will not permit her to turn round or step over the rope. She will usually accept the lamb straight away, but it may be necessary for the first few days to shorten the halter rope every few hours and stand by while the lamb is encouraged to suck.

Using birth fluids from a newborn lamb

To foster a lamb onto a ewe that is producing a single lamb takes rather more expertise. Firstly, be present at the birth. Immediately afterwards, before the ewe stands up or sees her own lamb, press a hand onto the abdomen just in front of the udder to see whether there is another lamb in the uterus. If there is, it can be felt as a firm, solid mass and as an additional check put one finger into the vagina to feel for any lamb about to be born. If no other lamb is present, keep the ewe in a lying position and if necessary wrap her head in a cloth so that she is blindfolded. She will then lie quietly.

Collect the lamb to be fostered, and using string, tie a front leg to the diagonally opposite back leg so that it is unable to stand. The foster lamb will usually be dry, but if it too is newborn and still wet, rub it briskly with a dry towel to dry it as much as possible. Use all the birth fluids available from the newly lambed ewe and her lamb, if necessary squeezing fluid from the legs and nose of the lamb, to saturate the foster lamb. Place the foster lamb at the ewe's head and remove the blindfold. Take away the ewe's own newborn lamb for about twenty minutes while the ewe licks the foster lamb, before returning it to its mother. The attempts of the foster lamb to stand while its legs are tied simulate the efforts of the newborn lamb trying to get to its feet. Wait another twenty minutes before untying the legs. During this time the newborn lamb, if it is healthy and normal, will have got to its feet and taken its first feed of colostrum from the ewe. Keep the ewe and lambs in a small pen for a few days to check that all is well and that both lambs are thriving.

Lamb adopters

These are boxes or crates which constrain a ewe in such a way that she cannot see or smell the lamb which is being fostered. The lamb has access to the ewe at all times and also has a safe area where it cannot be trampled on or kicked and butted by the ewe. The apparatus can be purchased or made up at home using plywood. It consists of a stall approximately 1 metre wide and 1.2 metres long, with a slot at the front just wide enough to take the ewe's neck comfortably, but narrow enough to prevent her from withdrawing her head when placed in it. A rail on either side of the ewe prevents sideways movement and provides the lamb with a safe, draughtproof area.

The theory is that after the milk has passed through the lamb for a few days, the lamb will take on the smell of the ewe and she will eventually recognise it and accept it as her own. This method is more successful with ewes that have only recently lambed. After release keep both ewe and lamb in a small pen until it is seen that all is well.

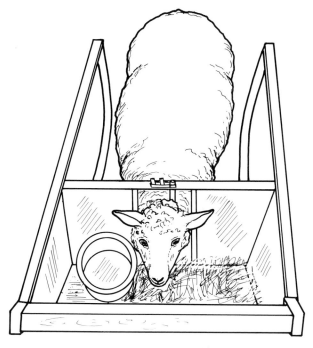

A LAMB ADOPTER

Ewes which are normally placid creatures can behave violently towards lambs which are not their own. Sometimes a ewe which has given birth to twins will reject one for no apparent reason. If both lambs are strong and active, tie the ewe to a corner of the pen using a halter, and after a few days she will usually accept both lambs. Try taking both lambs away for about half an hour. When they are returned she will usually be so pleased to see them that she allows both of them to suckle her.

After tailing, castrating and any fostering have been successfully completed, several ewes with their lambs can be grouped together in a small enclosure. Use a stock marker aerosol to identify lambs with their mothers by marking them on both sides of

the body with a similar number or symbol.

From now on, given an adequate supply of milk and grass, the lambs should thrive. Dose with a broad-spectrum anthelmintic at six weeks of age and again at monthly intervals to keep parasitic worms under control. If supplementary feeding is needed for the lambs, it can be started when they are around ten days of age. It can be in the form of pellets specially formulated for lambs or as a coarse mix containing ingredients which the lambs find particularly attractive — a kind of lamb muesli! Both will be available from your feed merchant. From an early age the lambs will also eat hay which, as is the case for all classes of sheep, must be of good quality.

The feed for the lambs should be made available to them via a lamb creep through which the ewes cannot pass. It should be

A thriving lamb

offered to them ad lib until they are eating around 300g a day. After that they should have a single feed each morning, the amount being dictated by the quality and the quantity of the grazing and the condition of the lambs. The feed, particularly when offered ad lib, should be protected from the weather and the lambs prevented from standing in the troughs when feeding.

Clean water should always be available to both ewes and lambs.

All the sheep are at risk of flystrike between April and November when they should be checked at least twice a day (see blowfly strike in Chapter 8). Short-term protection can be provided by dipping in an insecticidal dip.

If lambs are intended for the meat market, periodic weighing is required. With small numbers this can be done using bathroom scales or a sling or container and spring balance. The optimum weight for marketing is between 32 and 40kg liveweight, which will give a carcase of

approximately 48 per cent of that weight, depending on the degree of finish or fatness. Lambs from the smaller breeds of sheep and some of the rare breeds will be ready for market at lighter weights.

HAND REARING ORPHAN AND SURPLUS LAMBS

Day 1   It is essential that new-born lambs receive a minimum of one litre of colostrum or colostrum replacer during the first twenty-four hours of life (see page 55). It is important to thoroughly clean and sterilise all the utensils used in rearing orphan lambs.

Days 2 to 9   Introduce a good quality ewe milk replacer of preferably 30 per cent fat content, mixed with water in accordance with the manufacturer's instructions. Feed warm in four feeds to a total of one litre per lamb per day, using a bottle with teat attached.

Bottle feeding

It is possible to rear lambs on ewe milk substitute fed cold. It should be available ad lib and should be offered via teats fixed to the side of the pen, connected by plastic tubing to a supply of milk outside the pen. It is very important that the teats when fixed in position are above the level of the milk substitute in the container. The lambs will need to be taught how to use this feeding method by holding them onto the teat for the first few feeds. An infra-red lamp suspended not less than 1 to 2 metres above the pen will be appreciated by them.

Days 10 to 19   Introduce hay and concentrate feeding. As always, the hay should be of top quality and in order to tempt the lambs to eat concentrates, offer a small amount of flaked maize initially. Small pellets of compound feed purchased from your feed merchant can be introduced gradually. Restrict warm milk to two feeds per day.

Days 20 to 42   Continue with hay and concentrate feeding. Provide clean water. The water container and any feed troughs used should be scrupulously clean and set up in such a way that the lambs cannot foul them with their feet or with urine or dung.

Day 43 onwards   The lambs can be weaned and turned out onto clean grass. Feed concentrates twice daily at the rate of 250g per lamb per feed. They will be ready for marketing when they reach 32 to 40kg. The smaller breeds may be ready before they reach this weight.

CHAPTER 8   Some Common Diseases Affecting Sheep

It is always alarming to read about the numerous diseases to which sheep, like any other animal, can succumb, but good management, sensible fencing, anticipation of potential troubles and quick reaction will ensure that disasters rarely happen.

Blowfly Strike

This is probably the most common and troublesome of all diseases during the warmer months from April to November. The green or blue blowflies, which are slightly larger than houseflies, actively seek sites where they can lay their eggs and where the hatching maggots can obtain food. The fly finds the sheep particularly attractive when there are faeces contaminating the tail area, but even without this the natural smell of the sheep is sufficient to entice the fly to lay its eggs on the wool, and any open wounds can suffer the same fate. The eggs hatch in a few hours and the maggots start feeding by burrowing into the skin. The debris they create attracts more flies, and within a very short time the area affected rapidly increases in size.

However, long before this happens, the behaviour of the sheep will indicate to the observant shepherd that it requires attention. Sheep which have newly hatched maggots will be irritated by them and will waggle their tails vigorously. They may stamp the ground with a hind foot and twist the whole body from side to side while holding the head high in the air. The sheep must be caught and

SHEEP SHOWING SIGNS OF FLYSTRIKE

examined carefully. Apply a combined insecticide and antiseptic to the site of the maggots and also the surrounding wool to catch any that attempt to escape, rubbing the medication into the fleece with the fingertips.

If the strike is left untreated, the next indication will be a dark stain on the surface of the wool which is caused by the excreta from the maggots. By now the sheep will be feeling very sick and may be lying by itself under a tree or hedge. If moved it will stand with its head lowered and quickly lie down again.

74

In July and August the blowfly often lays eggs on the back of the sheep near the shoulders. When the maggots develop there, the sheep will stand with its head lowered and appear to be very sick. The tail waggling and body twisting are absent and one of the few visual signs is the tell-tale dark stain starting on the back and running down on to the shoulders. This appears after the maggots have been feeding for a day or two. Try to recognise the situation and deal with it before this.

Any sheep which have footrot are also at risk, as the distinctive smell of footrot attracts the blowfly to lay it eggs, and the resulting maggots penetrate the under-run hoof. Trim off the under-run sole and saturate with an insecticidal and antiseptic dressing.

Adult sheep are at risk of flystrike from late April until they are shorn. They are then free from infestation until the wool has grown to around 20cm, after which they can be protected for a few weeks by using a pour-on preparation such as Vetrazin. Lambs, unless they are shorn, are at risk all through the summer, but they too can be protected by using a pour-on preparation.

Blowfly strike is less likely to occur if preventative measures are taken routinely. Wool contaminated after lambing or as a consequence of diarrhoea should be cut off. Shearing is usually carried out in June but in exceptionally mild weather in the south of England it can be brought forward to the end of April. Clean, freshly shorn sheep are unlikely to suffer flystrike.

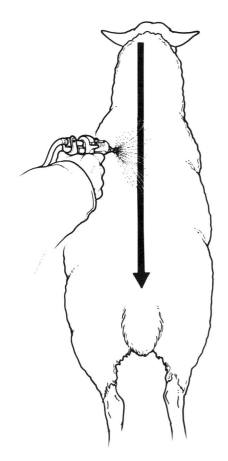

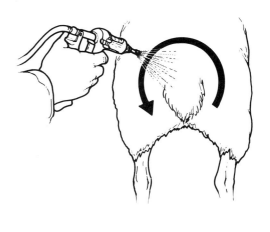

PREVENTING FLYSTRIKE

Sheep Scab

Sheep scab is caused by a mite which burrows into the skin and causes intense itching, along with rapid loss of wool and condition, and it requires immediate attention. If an infection is suspected, contact the vet immediately. He will be able to suggest the most effective treatment.

Scab is contagious and will rapidly spread through the flock if untreated. It is still a notifiable disease and unfortunately has become much more prevalent since annual dipping of sheep ceased to be compulsory. The disease will get into the flock only if there is very close contact with infected sheep or recently contaminated equipment, so it is wise to isolate, and if necessary treat, all sheep brought in before allowing them to mix with the flock.

Orf (Contagious Pustular Dermatitis)

Orf is a very painful condition affecting sheep of all ages. It is caused by a virus which enters the skin through minor cuts and abrasions, and it can also affect humans who come into contact with infected sheep, causing painful, scabby areas on the hands, wrists, inner arms and face. For this reason it is necessary to wear protective gloves at all times when dealing with this disease.

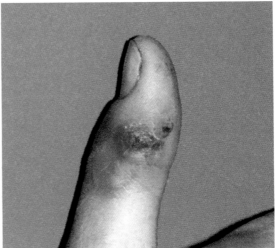

Orf on a finger

Having penetrated the skin, the virus multiplies rapidly, causing small spots which suppurate and join together to form grape-like clusters. It is confined to the non-woolly areas in both ewes and lambs. In lactating ewes, the udders and teats become infected, and in young lambs it affects the lips, gums and nose, resulting in the doubly distressing situation where hungry lambs with painful mouths are prevented from feeding because of the soreness and pain in the ewes' teats. The condition clears up in a few weeks and healing occurs. Orf can also affect older weaned lambs in the autumn and the genital areas of ewes and rams, making the rams very reluctant to work. The condition known

Sheep dipping

as strawberry footrot which causes raw red bleeding areas on the non-woolly areas of the faces and legs is believed to be related to the orf virus.

Being a virus, orf does not respond to antibiotics and is virtually impossible to cure. Secondary infections of the lesions can be guarded against by the use of an antibiotic spray.

It is possible to vaccinate against orf. The vaccine is applied to an area of bare skin, preferably between a foreleg and the chest wall. Do not use disinfectant or spirit to clean the skin before applying the vaccine, which is a modified form of the living virus, as this would destroy it before it can produce the desired effect. The skin must be lightly scratched where the vaccine is applied, and later a scab will form.

The disease is most troublesome in lactating ewes with new-born lambs, and flocks in which orf has occurred previously should be vaccinated two months before protection is required. On no account should sheep in flocks where the disease has never occurred be vaccinated. This is because the mild infection introduced by the modified living virus produces scabs at the site of the vaccination and these, as they fall from the sheep, are liable to affect any 'clean' sheep bought in.

An outbreak of orf can be controlled to a certain extent, especially in young suckling lambs, by vaccinating non-affected members of the flock. Seek the advice of your vet if an outbreak occurs, and follow the manufacturer's instructions carefully when using the vaccine.

Lameness

Lameness can occur in a number of different ways. Watch the sheep to see how it moves and which is the lame leg or legs. Catch the sheep, immobilise it and examine each foot in turn.

One of the commonest forms of

An extreme case of orf

lameness is caused by vegetable matter and mud becoming impacted between the hooves, and by mud becoming trapped in the layer of horn on the hoof wall of overgrown feet. Once this is removed and the hoof pared, the sheep will quickly become sound.

Look for a reddened and moist area between the digits and also for abscesses around the top of the hoof. If these are not found, feel the leg for any hot spots or swelling which might indicate a sprain or hairline fracture. Check that there are no dislocated joints.

Footrot and Scald

Footrot is a very painful and debilitating disease if allowed to go untreated and is directly responsible for causing huge financial losses to the sheep industry every year. While it is not a killer disease in itself, it can severely affect the performance of all classes of sheep from young lambs and ewes to rams, the latter especially at mating time.

It is caused by the joint action of several types of organisms, one of which is quite common in some areas of the UK. It causes scald — an inflammation of the skin between the two digits of the foot — and it can occur in all classes of sheep at any time of the year, but is more commonly seen when the sheep, particularly lambs, are grazing long, lush grass in the spring, and also shortly after ewes have been housed on straw bedding in the winter. The particular organism which causes scald can only live for three weeks away from the sheep, and careful use of a footbath can eradicate it from the flock, especially if clean land is available for grazing after footbathing.

True footrot occurs when a second organism invades the inflamed tissue between the digits. This causes an under-running of the sole of the foot, and eventually the horny outer wall, reaching right up to the toe. If this under-run tissue is lifted or pared off, the exposed area resembles the gills of a mushroom, except that it is usually dark grey in colour, and has a characteristic pungent, unpleasant smell. Treatment consists of paring away all the under-run horn from the sole and outer surface of the digit. The bulk of the material can be cut away with straight-bladed secateurs, but a very sharp knife should be used for the more precise trimming necessary to expose the fullest extent of the infection. The dividing line between healthy and infected tissue is a very fine one, and every care should be taken to avoid the bleeding which will occur if the healthy tissue is cut, especially in the region of the toe.

Here it should be stated that if footrot is treated as soon as a sheep is seen to be walking lame, a lot of pain for the sheep and a lot of work for the shepherd can be avoided. An antibiotic spray can be used to treat the infection, and the sheep should be stood on a hard surface while the foot dries.

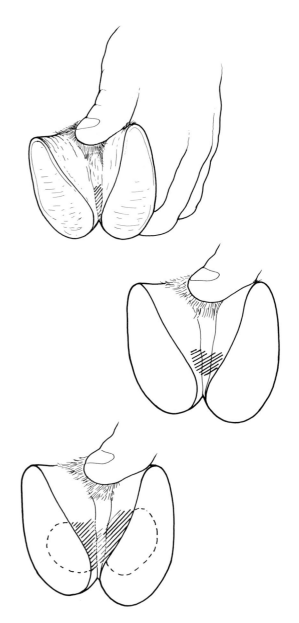

HATCHED AREAS SHOW
THE SPREAD OF FOOTROT

When footrot or scald is present, it is much easier to control if all the sheep can be stood in a shallow bath of ten per cent zinc sulphate solution for up to an hour. This long period enables the solution to penetrate pockets where the causal organisms might have been missed when paring. The feet should then be allowed to dry before returning to pasture. Formalin is also used for treatment and control, but it has a number of disadvantages. The fumes given off are toxic, and it is ineffective when the footbath becomes contaminated with dung or soil. It makes the hoof very hard and is very painful if used on feet where footrot has already been exposed. The feet should be soaked for a much shorter time with formalin solution, about two minutes usually being sufficient.

Both zinc sulphate and formalin are difficult to dispose of and both are poisonous to fish if allowed into streams, ponds or ditches. Zinc sulphate may be used and reused as it does not deteriorate in contact with vegetable matter.

When treating footrot it is important that all foot parings are collected and burnt and that all tools are disinfected by immersing them in ten per cent zinc sulphate solution between each sheep.

In severe cases of footrot, and particularly with valuable rams, consult a veterinary surgeon about the possibility of an antibiotic injection as treatment after paring. Prevention can be assisted by using a footrot vaccine. Two vaccinations are required initially, followed by two booster injections each year. The vaccine is expensive to buy and any partly used bottles must be discarded which adds to the expense for small flock owners.

Foot Abscesses

These can cause very painful lameness, occurring where the horn of the hoof joins the skin of the coronet. Consult your veterinary surgeon.

Granulomas or Proudflesh

When the feet are allowed to become badly overgrown, the horn may aggravate the soft tissue between the digits, or the under-run sole may rub against the sensitive layer below it and cause excessive growth with associated bleeding, especially from the area of the toe. This is a very painful condition and in the spring and summer may attract the blowfly to lay its eggs in the foot. (This is also the case with ordinary footrot.) Granulomas need to be surgically removed by a vet and even so they take a long time to heal.

Roundworms

All sheep carry some worms but it is the balance between numbers and the state of the sheep's resistance which determines whether disease develops. Pasture management is an important factor in controlling roundworm population. The old saying 'A sheep's worst enemy is another sheep' indicates that heavy stocking rates, coupled with infrequent moves onto clean pasture, can result in heavy infestation, with a resultant loss in condition and a reduced growth rate.

Roundworms occur in a variety of forms. Those that cause problems for sheep start life as eggs passed out with the faeces onto the pasture. Given suitable climatic conditions of warmth and moisture, the eggs develop into larvae which go through several stages before migrating onto grass stems where they are eaten by sheep. Two weeks later they are mature adults, mating and producing eggs.

This is an oversimplification of the matter because different types of roundworm prefer different parts of the digestive tract in which to live and produce eggs. Some can lie dormant in the gut until favourable conditions for further development come along. When this happens, a rapid explosion of activity occurs. Some types hibernate through the winter as larval stages on the pasture and only when the temperature rises in the spring (which also stimulates the growth of new grass) do they hatch in massive numbers to be eaten by young lambs and cause a lot of problems.

Diarrhoea, dehydration and reduced appetite are symptoms which can indicate worm infestation, although these can be confused with symptoms of other conditions. Even with moderate or light infestations there will be a reduced growth rate in the lambs and the fleeces will appear to be dull, dry and lacking in bloom.

Anthelmintics are drugs that control internal parasites and a wide range of them is available for the control of roundworms. For the treatment to be effective, it is very important to follow the manufacturer's instructions carefully to match the dosage required to the weight of the animal being treated. There is evidence that some types of roundworm are becoming resistant to some of the drugs being used. Check with your vet as to the most suitable anthelmintic if your sheep have a particular problem.

Clean pasture is invaluable in providing an efficient management aid in that newly dosed sheep can graze without the challenge of immediate re-infestation. Similarly, mixed grazing with cattle can do a lot to reduce the uptake of worm larvae.

Tapeworms

Tapeworms are flat, ribbon-like parasites which live in the small intestine of the sheep. The worm consists of a tiny head which attaches itself to the wall of the gut, either by a sucker or with hooks, and it feeds in that position. The ribbon-like body consists of a series of segments which are grown from the head and which may reach several metres in length. They are in fact packets of eggs which are shed from the end of the worm and pass out with the faeces. They require an intermediate host for their further development, and can be easily seen as flat, white, short pieces of tape. Although unsightly, they rarely cause any great harm

to the sheep, unless present in large numbers.

Many anthelmintics used to control roundworms are also effective with tapeworms. Details will be found on the manufacturer's package.

Gid

There are a number of different kinds of tapeworm, and this disease of the central nervous system is caused when sheep eat grass contaminated by the larval stage of a tapeworm found in dogs. Over a period of time the larvae migrate through the bloodstream and begin developing when they reach the nervous tissue, including the brain. Here they grow into a cyst filled with fluid containing tapeworm heads.

The symptoms which may indicate gid can vary, and are easily confused with other nervous diseases. Sometimes the affected sheep will lie with the head turned to one side, or hold it abnormally high. Others will walk in circles, always turning the same way, and all will eventually go down, become comatose and die. The time taken for this to happen varies considerably, and if the veterinary surgeon is called in early and diagnoses gid, the chances are that he will be able to operate successfully and remove the cyst.

If the uncooked head of a dead sheep suffering from gid is eaten by a dog, the tapeworm heads develop into adult tapeworms and the life cycle of the parasite begins again. It is possible to break this cycle by ensuring that any sheep offal fed to dogs is boiled for at least one hour. All dogs owned should be dosed with a medication obtainable from your vet, which is capable of killing all of these tapeworms, including the heads. This should be done every two months, but of course it will not prevent other dogs and foxes coming onto the property and contaminating the grassland.

Hydatid Disease

There is another type of tapeworm, less than a centimetre in length, which lives in the small intestine of the dog. The ripe segments from these tiny tapeworms are passed out in the faeces of the dog, some adhering to the hairy coat. Infection of the sheep occurs when the grazing animal eats grass contaminated by infected dogs. Humans, especially children, can become infected by stroking and cuddling dogs and then accidentally swallowing the tapeworm eggs.

The hydatid cyst will grow in any part of the body, but commonly in the lungs and liver. Human beings cannot become infected by eating lamb or mutton. Prevention by strict hygiene when handling dogs and by regularly dosing them with an anthelmintic against this particular tapeworm, called *Echinococcus granulosus,* is desirable. Never feed raw sheep meat to dogs — it should be cooked for at least an hour — and dispose of any carcases so that they cannot be scavenged by dogs or foxes. Discuss prevention with your vet.

Liver Fluke

The fluke is a flat, leaf-like parasite which is common in wet, marshy areas. It requires an intermediate host, a mud snail, for its development. For the sheep, the problem starts when it eats grass which is carrying the encysted stage of the fluke. On reaching the sheep's gut, the immature fluke makes its way through the gut wall and searches out the liver. Once in the liver it feeds voraciously, and on becoming adult it anchors itself in the bile duct and takes blood from the wall. It lays eggs which pass out with the faeces. These then hatch and seek out a mud snail, in which they multiply, thereafter leaving the snail and encysting themselves on grass to begin the cycle again.

Affected sheep become very anaemic and lose condition. A watery swelling appears under the lower jaw and the membranes of the eyes and mouth become very pale.

If the presence of fluke is suspected, take a dung sample to your vet who will be able to tell you whether fluke eggs are present. The vet will also advise on the method of treatment most pertinent to the condition of your sheep. Cattle, goats, deer and rabbits can also act as hosts and carriers of fluke.

Pulpy Kidney and Other Causes of Sudden Death

There is a group of diseases (including tetanus) caused by *Clostridium* bacteria which can cause sudden death, and it is very nearly always the best sheep which are affected. Fortunately there are very good vaccines available which will give complete protection from specific diseases if given routinely and according to the manufacturer's instructions. It is recommended that for all breeding stock a '7 in 1' type vaccine is used. If well grown and sold for slaughter before 16 weeks of age, lambs born in a vaccinated flock will not normally need vaccination, as the protection given by antibodies in the ewe's colostrum will be sufficient. Other vaccines giving protection against fewer clostridial diseases are available.

It should always be assumed that any bought-in sheep are unvaccinated. They should be given a full dose of vaccine immediately, followed by a booster dose four weeks later. Another booster dose for the ewes four weeks before lambing commences ensures that newborn lambs are protected, provided they get an adequate supply of colostrum. If lambing is protracted it may be necessary to give a further booster dose of vaccine to late-lambing ewes.

Home-bred ewe lambs born to vaccinated ewes and intended as flock replacements will require a primary vaccination at two to three months of age, followed by a secondary dose four to six weeks later. If they are bred from in their first year they, like the ewes, will require a booster dose four weeks before lambing. Similarly, ram lambs kept for breeding should be treated like the ewe lambs, but in subsequent years a booster dose around four weeks before mating commences will suffice.

This vaccine should be given subcutaneously at a site high on the neck of the sheep. Occasionally an abscess or small nodule will form at the injection site and if this is on the neck, it will not reduce the value of the skin after slaughter.

Pneumonia

The bacteria which cause pneumonia in sheep come in many different types. Deaths from pneumonia can occur in all classes of sheep and at any time of the year. Often the first sign of the disease is a dead sheep, or a short period of laboured breathing and frothing at the mouth followed by death. Other forms of the disease can be treated with long-acting antibiotics if detected early.

Pasteurellosis (pneumonia) is a very complex and variable condition. Periods of stress, which can include bad weather, can predispose a flock to infection, and if the level of stress is great enough outbreaks can occur even though the sheep have been vaccinated.

A post-mortem examination of sheep which die from pneumonia will reveal plum-coloured lungs surrounded by bloodstained fluid, and in many cases death will have occurred so rapidly that it is impossible to observe any symptoms.

It is possible to vaccinate against certain types of pasteurella, and to be effective the vaccine must be appropriate to the particular requirements of the flock as determined by laboratory tests. It is possible to buy a combined clostridial and pasteurella vaccine, but little, if any, immunity to pasteurella is passed to the lambs in the antibodies

contained in the colostrum, compared to the cover given against the clostridial diseases. If early protection is required for the lambs, they must be given pasteurella-only vaccine when two weeks old if they are born to ewes which have previously been vaccinated.

Lambs born to ewes which have not previously been vaccinated can be given pasteurella-only vaccine at birth, followed two weeks later by a second injection. Ewe lambs retained for breeding will require one annual booster injection.

METABOLIC DISEASES

Metabolic diseases associated with feeding and production are sudden in onset and can lead to death in a very short time. They arise as a consequence of food intake not meeting urgent requirements.

Pregnancy Toxaemia

Also known as twin lamb disease, this occurs during the last month of pregnancy, most commonly in ewes that are carrying two or three lambs. At this time the unborn lambs are making heavy nutritional demands on the ewe. Pregnancy toxaemia can affect both thin and fat ewes and is usually caused by an insufficient uptake of nutritious food.

Concentrates should be fed during the last six weeks of pregnancy, starting with 100g per day, increasing weekly to reach 750g per day by lambing. The protein content of the concentrates should be seventeen per cent, and it is helpful if the merchant supplying the feed gives you a list of the ingredients so that the stated analysis can be checked.

Ewes affected by pregnancy toxaemia get separated from the rest of the flock, refuse food and become blind. After a few days they are unable to stand and they assume unnatural positional behaviour. Front legs may be stretched out in front and the head pressed down into the ground. Symptoms include twitching and trembling of the lips and ears, and grinding teeth. If the ewe is left untreated, death follows fairly quickly. The advice of the vet should be quickly sought if pregnancy toxaemia is suspected, and the ewe moved to warm, comfortable conditions under cover.

Prevention of this disease is easier if sheep are housed, as a more careful check can be kept on their condition. The target condition score at lambing is 3.5, and thin ewes and shy feeders which are often bullied at the trough should be separated and given extra concentrates. Ewes which are lame or have teeth problems are also at risk, as are ewes kept outside in periods of very bad weather. See that all the ewes have sufficient trough space. It is helpful to be able to put all the feed in the troughs before allowing the sheep access to it, perhaps in a separate enclosure or an adjoining field. This will allow fair shares for all.

The bulk part of the diet should be of good quality hay, and the concentrate fed first. A stomach full of poor quality food is of no use to a ewe carrying three lambs. Four-fifths of the eventual birthweight is gained during the last two months of pregnancy.

Overfat ewes too are at risk, simply because internal layers of fat restrict the space available for any increase in nutritious food required to feed the unborn lambs.

Death is caused by a breakdown of the metabolism and a lowering of blood sugar levels. One school of thought suggests that if the ewe is shorn the metabolism is stirred into action again by the demands of the ewe as she attempts to convert energy into warmth!

It is best to prevent the disease by careful management of the flock, particularly in the later stages of pregnancy, as the recovery rate is very low.

Hypocalcaemia

This disease can occur during late pregnancy or early lactation, and it is caused by a low level of calcium in the blood. Listlessness and depression precede loss of consciousness, but a subcutaneous injection of 100ml of twenty per cent calcium borogluconate will effect a rapid return to normal. Divide the dose and inject at five or six different sites on the body.

Hypomagnesaemia

The onset of this disease can be extremely rapid and usually occurs when newly lambed ewes are turned out onto very lush grass. It is caused by low levels of magnesium in the blood, and is characterised by a stiffness when walking, and twitching and spasms, followed by convulsions and death. Prompt treatment, consisting of a subcutaneous injection of 100ml of magnesium-fortified twenty per cent calcium borogluconate, divided and injected at several sites as with hypocalcaemia, is necessary.

The main problem in treating this disease is the speed at which it develops, and unless the person in charge is present and very observant, quite often the only indication is a dead sheep. Prevention can be assisted by offering a mineral mixture containing calcined magnesite, but as this is an unpalatable ingredient it needs to be disguised to make it acceptable by adding some molasses or treacle. It is possible to buy compound feeds containing calcined magnesite, and these should be introduced into the rations gradually just before lambing, and continued for a few weeks afterwards while animals are at risk.

Heavy dressings of nitrogenous fertilisers encourage a rapid growth of young grass in the spring which also reduces the uptake of both calcium and magnesium, a point to be borne in mind when considering how best to manage the available grassland.

DISEASES ASSOCIATED WITH LAMBING

Abortion

There are a number of different organisms which can cause abortion in sheep. Abortion can also be caused by other factors such as worrying dogs, malnutrition, incorrect handling and being forced too quickly through narrow gateways, to name but a few. In some cases of abortion early or midway through pregnancy the foetus may be so small that it passes unnoticed by the shepherd. A single ewe aborting may give no cause for alarm and be dismissed as one of the accepted risks of keeping livestock.

When a number of abortions occur, producing dead or very weakly foetuses, the vet should be called in without delay as it is only by microscopic bacteriological examination of aborted material that an accurate diagnosis can be made, and, if appropriate, the type of vaccine required to minimise further losses be recommended.

Domestic cats are responsible for spreading the infection causing one type of abortion in sheep, which is also contagious to humans. Consequently, extreme care should be taken when handling and disposing of any aborted material, and pregnant women should stay away from ewes at lambing time.

Mastitis

Mastitis can be present in the ewe flock in both chronic and acute forms. The chronic form is usually discovered when the ewes are being inspected prior to selection for the breeding flock. It is identifiable as a fibrous lump or series of lumps in the udder. If the ewe is used for breeding, the affected side of the udder will not produce any milk. It is not known when infection occurs and there is no known cure. Affected ewes should be culled.

The acute form commonly occurs early in lactation. Affected ewes are clearly ill and

Mastitis

appear to be walking stiffly, and the udder is swollen and grows increasingly cold to the touch. The udder rapidly becomes inflamed, turning red or purplish-blue and eventually black. If the teat is drawn, a bloodstained fluid comes away. The inflammation may extend along the belly of the ewe if she survives and eventually all the dead coloured tissue will fall off leaving a raw, bleeding surface. Healing takes a long time and the exposed tissue is vulnerable to flystrike. Consult the vet as to the best course of action. There is no effective treatment for this condition, and if the ewe survives she should be culled.

Navel Infections

The umbilical cord of a newborn lamb is moist and immediately after birth it is in contact with many different types of bacteria on the ground. This is the route by which certain harmful organisms gain entry to the body of the lamb. It is comparatively easy to prevent this from happening by applying an antiseptic dressing to the navel as soon as the lamb is born and then placing the lamb on a bed of clean straw. The aim is to dry up the cord as quickly as possible and to disinfect the area of the body around it.

Several preparations are suitable for use as a dressing:
- Tincture of iodine
- Antibiotic aerosol
- A weak solution of copper sulphate

A dressing which can be contained in a wide-necked bottle is preferable to an aerosol as the wet cord and the area immediately surrounding it can be totally immersed in the liquid. Hold the lamb vertically by its forelegs and put the navel cord into the neck of the bottle. Lift the bottle at a right angle to the lamb and press it gently into the belly.

Coccidiosis

Coccidiosis is a disease of young lambs. Older sheep will have developed an immunity through previous contact with the parasite, but they can still pass on the infection. The disease occurs mostly in housed flocks, although it can also infect lambs on heavily stocked pasture through contaminated faeces.

Infected lambs become unthrifty with a loss of appetite and develop a mild diarrhoea which causes dehydration. If coccidiosis is suspected, consult your vet over treatment and preventative measures to be taken in subsequent years.

NOTIFIABLE DISEASES

There are a number of important diseases which can affect sheep and which, if suspected, must be reported by the owner or person in charge of the animals to the police, or a veterinary officer of MAFF in England or the Department of Agriculture in Scotland. These are called notifiable diseases and a full list can be obtained from Animal Health Division A, Hook Rise South, Tolworth, Surrey KT6 7NF.

Anthrax, foot and mouth disease, rabies and sheep pox are the ones on the list that are most likely to affect sheep.

ZOONOSES

Animal diseases which are communicable to human beings are called zoonoses, and while some may cause only slight problems, others can be dangerous and even life-threatening. Anyone who is working with sheep or in contact with them when a zoonotic disease is present or suspected should seek medical advice without delay if they feel unwell.

The following are zoonoses:
- anthrax
- rabies
- salmonellosis
- hydatid disease
- orf
- brucellosis
- chlamydiosis
- listeriosis
- toxoplasmosis
- all abortion agents

The last five on this list all pose a threat to pregnant women, as they can cause abortion or serious defects if in contact with the unborn foetus. Pregnant women should keep away from sheep at lambing time.

YOUR VETERINARY SURGEON

Some veterinary surgeons are more involved in sheep than others, so try to find and get to know one with whom you can readily communicate. Also, do this before you encounter an emergency. It will cost you money, but a general advisory visit to discuss management, feeding, a vaccination programme, a worming policy, etc., could prove to be a good investment.

Don't forget that you can take a sheep which is in trouble to the surgery and save yourself a call-out fee.

CHAPTER 9 Keeping Records

All flock owners will find it extremely useful to keep a record of day-to-day events with the flock. The records can contain as little or as much of the history of the flock as the owner decides, and can include details relating to individual sheep, food purchased, grassland management, veterinary treatment, etc. Records of the performance of individual sheep start with accurate identification. Contact your Trading Standards Officer and MAFF to keep up to date with your legal obligations.

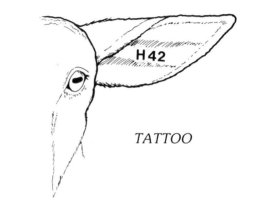

TATTOO

IDENTIFICATION

Each sheep in the flock should carry a permanent mark which will identify it and distinguish it from its flockmates in all the various forms of recording. In addition to an individual number, the mark can also indicate the year of birth and in some cases the sire.

Permanent identification can be one or more of several different kinds. It may consist of a tattoo mark in the ear, a metal or plastic tag in the ear, or a system of notches, indicating numbers, made around the outer edge of the ears with special pliers. Sheep with horns can be heat-branded, but it must be remembered that horns can be lost through accidents and fighting and the brand can become difficult to read in older sheep.

Tattoos are easier to read in the ears of white-faced sheep as they show up clearly when black tattooing ink is used. For sheep with coloured faces white or green tattooing ink is more suitable, but in all cases it is essential that the mark is placed centrally in the ear, between the two main sinews. With some pedigree breeds one ear is reserved for the flock number and the official

EAR TAG

registered mark of the breed society, while the other is left for the use of the flock owner.

Eartags come in a variety of shapes and sizes in both metal and plastic. Some metal tags are tamperproof in that they cannot be removed from the ear and re-used. The

An ear tag

type of plastic tag requires special pliers for application.

Ear notching, again using special pliers for the purpose, can be done at any age. It does not, however, indicate the year of birth, and for this a single letter tattooed in the ear is needed. The sheep's left ear is used for units and the right ear for tens. Five positions on the left ear will give sufficient spaces from one to nine, and from ten to ninety on the right. Sheep number seventy-nine would require notches denoting fifty and twenty on the right ear and five and four on the left, with a tattooed letter in the right ear indicating the year of birth.

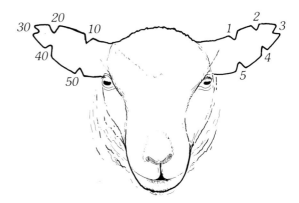

EAR NOTCHING

manufacturers of these tags engrave numbers on them which are recorded to the name of the purchaser, although other numbers and letters can be engraved at the purchaser's request. The manufacturer's instructions should be carefully followed when using these tags on young lambs, remembering that the ear will grow and require more space within the tag. Special pliers are needed for their application. An advantage of using plastic tags is that they come in a variety of colours which can be seen at a distance. If a different colour is used for each year of birth, it is a great help in identifying a sheep without having to catch it. The tags also come in a variety of shapes and degrees of flexibility, but as they are all at risk from snagging on fences and equipment and can also on occasions break, it is advisable to put one in each ear. Each

Another method of identification is to use a temporary fleece brand. If this is to remain legible as the fleece grows, the size of the numbers should be fairly large. A metal rod of quarter-inch diameter is ideal for forming the numbers, which should measure 150mm form top to bottom. Short pieces of rod can be welded together to make up the number, and a handle welded on the back. These brands are used by dipping them into a shallow container of marking fluid, and positioning the number high up on the side of the sheep, rather than on the back. This will enable the number to be read from a distance, which is particularly useful at

FLEECE BRANDING

fleece. It is available in several different colours. The use of ordinary paint or tar will be heavily penalised by the British Wool Marketing Board when the fleece is graded, and the value of the whole fleece considerably reduced. This is because it is impossible to scour out these materials when the fleece is being processed.

SYSTEMS OF RECORDING

Once there is a system of identification, which is essential in pedigree flocks, records can be kept of the health and performance of the sheep.

Some breed societies have designed a pro-forma with headings under which details can be entered as they occur, and from which an accurate assessment can be made of the performance of the flock from year to year.

A loose-leaf system, based on one page for each sheep, can, for example, include details of the breed, the ear number, the age and a record of the sire and dam. In the case of a ewe, details of mating, the ram used, the date of service, together with the date of lambing and the sex and birth weight of the lambs born, any subsequent weighings, how disposed of and the price obtained can be recorded on a form like the one below.

For a ram, in addition to individual details, the number of ewes served, the total of lambs born to them, alive and dead,

lambing time when several people may be involved in looking after the sheep. It also avoids the necessity for catching individual sheep in order to identify them. The fleece brand need not necessarily be the same number as the ear number as long as a cross-reference list is kept. Thus the flock can be branded with consecutive numbers disregarding the individual numbers which will possibly contain a year letter and a lengthy figure.

Although curiously referred to as 'paint', the marking fluid that should be used is of a type approved by the British Wool Marketing Board as it can be scoured from the shorn

EWE RECORD CARD

BREED:		EAR No:		BORN:		SIRE:		DAM:	

Year	Ram Used	Date Lambed	Ear Nos & Sex of Lambs	Birth Weight	28 Day Weight	Disposal	Date	Wt. Price	Destination

together with details of the total sum realised when the lambs are sold, will allow a comparison to be made between the performance of the rams used, and between the performance of the rams used, and between the performance of the same ram in different years.

In addition to the performance of individuals within the flock, a record of the sheep food purchased and times in the year when feeding was considered necessary, can be recorded. Any cultivations required by the grazing land and fertiliser applied, together with capital costs incurred for housing, fencing and equipment may provide interesting (or frightening!) information and be used in calculating a profit and loss account.

A separate record of all medicines purchased and administered, including the dates, is a legal requirement. To assist in this a booklet entitled "Animal Health Record Book", which incorporates the Code of Practice for the safe use of veterinary medicines on farms, has been produced by the National Office of Animal Health. Copies of this booklet may be obtained from MAFF whose officers have the statutory right to inspect the records kept.

Medicinal Records

Under the Food Act 1990, records must be kept for any drug used which has a withdrawal period before the animal can be slaughtered for human consumption.

Disposal of Carcases

The disposal of carcases of any animals which die can be a problem. If it is impracticable to bury carcases at home, the vet will advise on other methods of disposal which may be available to the flock owner. The Animal By-products Order requires that the carcases of all animals that die must be sent to a licensed knacker or burnt or buried on the premises. Movement records are required for any carcase taken off the premises.

The Sheep & Goat Registration, Identification and Movement Order, 1996

There is a statutory requirement for all persons in charge of sheep to make a record of all movement onto or from the farm. A book for keeping this record can be obtained from the Animal Health department of the local authority, normally the County Council. If there is a major breakdown in animal health, this record is vital for tracing animals which might have been contacts at the source of the breakdown. The record of movement of animals must now be kept available for inspection for three years.

The Welfare of Animals Transport Order 1997, copies of which may be obtained from H M Stationery Office, 49 High Holborn, London WC1V 6HB, require that details of the transport of sheep be recorded within eighteen hours of completion of the journey. The time of loading and unloading and the time when the animals are fed and watered must be entered as they take place. The Welfare of Animals During Transport Order 1994 requires an Animal Transport Certificate in regard to certain things, e.g. length of journey, size of vehicle, ventilation required, safe internal structure, etc. Regulations came into force in 1996 affecting everyone who keeps sheep. These regulations are designed to provide information for checking claims for the Sheep Annual Premium and they will also enable any outbreaks of disease to be traced back to the farm of origin.

The main requirements of the order can be summarised as follows:

Registration Even if you own only one sheep you must register with your local Animal Health Office. You will then be given a registration number.

Animal Health Act 1981
The Sheep and Goats
(Registration, Identification and Movement)
Order 1996

SHEEP FLOCK RECORD

Date	Running total of female sheep which are over 12 months old or which have given birth	Number of sheep moved onto holding	Number of sheep moved off holding	Number which have not given birth reaching 12 months old	Number under 12 months old which have given birth	Number and cause of sheep loss

In addition, details of the obligatory Sheep and Goat Movement Document (obtainable from your Trading Standards Office) will need to be recorded

Movement All movements of sheep to and from your holding must be recorded within thirty-six hours of the movement taking place. The records must contain:

● the date of movement
● the identification marks of the animals
● the total number moved
● the holding where the animal was born
● the holding from which the animal is moved
● the holding to which the animal is moved

Identification As recommended earlier in this chapter, it can be a great help if all your sheep are identified individually. Under the new regulations, if the sheep are being exported the consignor must ensure that they carry an eartag or tattoo with UK and the flock registration mark. This tag can also be used to carry the sheep's individual mark. For movement within Great Britain the animals can carry the sheep's individual mark. For movement within Great Britain the animals can carry the export tag, or a management tag or tattoo, a paint mark, an ear notch, a horn brand or a leg band. The identification must remain visible for the duration of the journey, and the animals must be accompanied by a movement document which must contain the following details:

● the identification mark
● the address of the consignor's premises
● the date of movement
● the number of animals being moved
● the address to which the animals are being moved

The document should be given to the person receiving the animals and must be kept for six months. You do not need to mark your animals or complete a movement document if you are moving sheep between holdings in the same occupancy or to common grazing.

Records for Sheep Subsidies

If you claim the Sheep Annual Premium (see page 106), you must keep a continuous up-to-date record of all female sheep over twelve months of age, or under twelve months if they have given birth. The legal requirements are that you record the total number eligible for subsidy on January 1st each year, and this must be done by January 31st. You must also record changes in that total number within fourteen days of:

● sheep being moved onto or off the holding
● sheep which have not given birth but which have reached twelve months of age
● sheep under twelve months which have given birth to a lamb
● any lost sheep
  Your records must also contain
● the reasons for any change in the total number
● the causes of losses if known
● the date of every entry

These records must be retained for four years from the end of the year in which the last entry was made.

Note that no subsidy will be paid if no records are kept.

The above information has been taken from a draft leaflet circulated by MAFF. Some slight alteration may occur and for current information you should contact your local Animal Health Office.

TRANSPORT

There are occasions when it is necessary to transport members of the flock. It might be to a show, to a market or fair or to a dipping or shearing centre. There is a statutory obligation under the Animal Health Act 1981 and the Welfare of Animals

A COMMERCIAL TRAILER

Transport Order 1997 as amended for certain legal requirements to be observed when transporting stock.

The main ones apply to all forms of transport and may be summarised under the following headings.

The roof must be weatherproof.

Ventilation should be adequate and appropriate to the number of stock being transported.

The interior of the transport vehicle shall be free from projections likely to cause injury or suffering.

The floor shall be of non-slip design and/or strewn with straw or similar material to provide a proper foothold.

Ramps shall be designed to prevent slipping and have no gap between the ramp and the floor of the transport vehicle in which a foot or leg can be trapped.

Sidegates must be of sufficient height and strength to avoid falls from the ramp.

Partitions/dividers must be fitted if the floor exceeds 3.1m in length. If a full load is not carried, the pens will need to be smaller.

A home-made sheep trailer

CHAPTER 10

The Shepherd's Calendar

In keeping and caring for sheep in flocks large or small, there follows a sequence of events throughout the year. It is the same for both early- and late-lambing flocks and begins in late summer or early autumn with the selection of breeding stock. The future crop of lambs will depend to a certain extent on daily observations by the shepherd of the prevailing conditions. The state of the grazing, the weather, supplementary food and preventive medicine all need to be assessed constantly and adjusted as necessary.

July, August, September

Make a thorough inspection of the ewe flock, paying particular attention to teeth, feet and udders. Mark for culling any that do not meet basic requirements, segregate them and sell them for meat.

If additional sheep are required, buy the best that can be afforded.

October

Get the ewes onto a rising plane of nutrition by putting them onto clean, fresh grazing or by feeding a small amount (100g) of concentrates each morning. Aim for a condition score of 3.5 at mating.

Fit the ram's harness in position a few days before mating is to begin and tighten if necessary when adding the coloured crayon.

Introduce the ram five months before the planned date for lambing, and record the date on which each ewe is served. The average gestation period is 147 days.

November

Stop feeding concentrate to the ewes when they have been mated, but feed hay if the quality and quantity of the grazing deteriorates. The hay must be of good quality, free from mould, and sweet-smelling. It should be fed from covered hayracks as sheep will not eat wet hay.

December, January, February

Consider housing the ewes if the weather is bad, using barley straw for bedding. Continue feeding hay ad lib. Introduce concentrates six weeks before lambing is due to begin. Vaccinate each ewe with a booster dose of vaccine against clostridial diseases immediately before lambing.

Prepare the lambing kit and ensure that all records are up to date.

March, April

Lambing. Record lamb dates of birth and birth weights. Carry out tailing and castrating.

Continue feeding ewes, adding succulent foods such as roots, cabbage type green food or soaked sugar beet pulp if possible.

Creep feed lambs.

Give ewes a dose of broad spectrum anthelmintics and trim feet. Weather permitting, turn ewes and lambs out onto clean grazing by day, returning to a communal pen in the sheep house at night. After a period of hardening off they can be left out overnight.

Dose lambs against worms at six weeks of age and thereafter at monthly intervals.

Foot trimming

May, June

Reduce the concentrate feeding for the ewes as the quantity and quality of the grazing increases. Offer hay in covered hayracks if there is any scouring.

Carry out crutching, shearing and dipping against fly strike.

July

Wean lambs onto fresh, clean grazing. Sell lambs destined for the meat market as they reach the optimum weight.

Vaccinate any remaining lambs against clostridial diseases as they reach sixteen weeks of age.

August, September

Attend store lamb, breeding ewe and ram sales. Make up the breeding flock.

CHAPTER 11   **Showing**

Many small flock owners get pleasure from taking their sheep to shows and competing for the honour of winning prize cards and rosettes. Many local shows now have classes for all types of sheep and lambs, and as they are usually held at weekends it means that more people can spare the time to attend.

All sheep shown, whether pure or crossbred, should have good conformation and be well-grown for their age. Depending on the breed, they should be in good condition, with the down breeds having a condition score of at least 3.5 or 4. The fleece should be clean, the feet well trimmed and it is a great help for both sheep and shepherd if the sheep is taught to walk on a halter. Halter trained sheep are able to show themselves off to the judge without involving the handler in a rodeo routine!

Selection

When selecting a sheep for show, start looking well before the date of the show and select an animal that has the indefinable quality of 'character'. This is the sheep that is always appealing to the eye in any position and one that always stands out from its flock-mates. If it is a purebred or pedigree sheep, check the list of points which have been drawn up by the breed society concerned as being particularly desirable. Match the selected sheep as nearly as possible to that list. Select a sheep with a bold eye and an alert expression, a straight back and standing four square on straight legs. Observe the way it moves when walking towards and away from you, especially when being led on a halter.

Halter training

Some show judges ask for an entry to be turned loose in the show ring so that they can see for themselves how the sheep moves naturally, but others ask for it to be paraded on a halter. Use a cotton halter for training rather than one made from polypropylene and fit a piece of rubber tubing to the part of the halter which goes under the lower jaw. This will prevent the

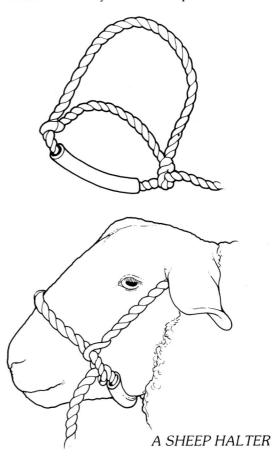

A SHEEP HALTER

Teaching to lead on a long rope

halter from becoming too tight if the sheep puts up a lot of resistance, and thus prevent chafing the skin which might lead to an unsightly scar. Have a long lead rope on the halter when beginning the training and walk behind the sheep in such a way that it can be restrained if it tries to break away. A push from behind will usually encourage it to walk forward. After a half-hour session each day for a few days, it will quickly become accustomed to the halter and the handler, and walk beside him in a natural way.

The next step is to get the animal to stand still while on the halter and to stand in a way in which it shows itself off to the best advantage. It may be necessary to move a leg slightly to achieve the best position. This can be done either by gently pushing or pulling the sheep to one side or the other, or by moving it with the hand.

Standing correctly

A mixed breed class

Showmanship

Wear a clean white coat and keep the sheep in a correct position all the time that it is before the judge. It should not be allowed to graze or turn around or lie down. Onlookers might be amused to see this, but it could cost a prize if it happened just when the judge looked back in order to make a comparison with another sheep. Keep the head of the sheep at a naturally comfortable height while it is being held, neither too high nor too low. Stand in front of the sheep while the judge is standing behind the line-up, but move to one side so that he can see the front when he comes around to the other side. Watch the judge as well as the sheep all the time and try to convince him by showmanship and without words that you have the best sheep!

A champion ram

Dressing

The fleeces carried by the various breeds of sheep vary a great deal between the fine, short staple of the Southdown and the long, coarse wool of the Scottish Blackface. Others like the Wensleydale have long, lustrous fleeces which hang in ringlets, and the Herdwick and Welsh Mountain contain a high proportion of kemp in their wool.

Different types of wool require different treatments before being presented for the judge's inspection. Tradition dictates that on some breeds unnatural shapes are created, using hand shears, while others have a minimum of trimming with a programme of dipping in a variety of dips. In all cases, the fleece should be clean and free from dung and vegetable matter.

Any washing that is required should be done about ten days before the show. Tie up the sheep and soak with water from a hosepipe or buckets. Use soap flakes and work up a lather in the wool, using the fingers to tease out any matted locks. If a swimbath is available, fill it with clean water and allow the sheep to swim in it to rinse out the soap and dirt. Keep the sheep on a halter while swimming, as fat sheep with heavy fleeces can be at risk in the bath. Use the halter to pull the sheep to the ramp leading out of the bath if the animal appears to be distressed in any way. If no swimbath is available for rinsing, use a hosepipe or bucket with plenty of clean water.

Choose a dry day for the washing as it is important that the sheep dries quickly. Never put the sheep under cover while still wet. Washing with soap will leave the fleece soft, white and open in texture, and it will respond to being dipped in a bath containing show dip. These dips are designed specifically to firm up the wool in shortwools and to add to the lustre and purl effect in longwools. They also take away the flat whiteness which is apparent after washing in soap.

As each breed of sheep has its own traditional shape or appearance when brought out for show, it is not proposed here to go into the requirements for each breed. Suffice it to say that if the plan is to produce show entries trimmed with hand shears, then take a course by an expert on how it should be done. The equipment required — hand shears, sharpening stone, carding wire on heavy wooden backs, halters and an old shearing comb for cleaning the carding wire — is modest, but the skill in using it is not!

Take consolation from the fact that a good sheep is attractive to the experienced eye whether it is trimmed or not.

Lambs can be shown in their natural fleece, while older sheep will look better if they are carrying about four months' growth of wool. All that is required on a sheep that has been docked is that the wool on the end of the tail is squared off parallel to the ground.

The Texel breed has stipulated that all its show animals must be shown in a natural condition. Any trimming or shaping is not allowed, although faces and legs can be washed clean.

Fleece wool

A less demanding type of showing is in classes for shorn fleeces. Select the fleeces while they are still on the sheep and swim the sheep in clean water without soap, dip or disinfectant added, around four weeks before shearing. Ask the shearer to take special care when shearing and wrap the shorn fleece according to the instructions in the show schedule. Place the fleece in a large paper sack as this will help it to retain its wrapped shape. Avoid plastic sacks which will cause the wool to sweat. Identify your fleeces with a clean well-written label.

ROLLING A SHORTWOOL FLEECE

A

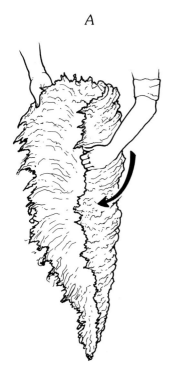

Place the flesh side of the fleece down and fold flank towards the centre.

B

Do the same with the other side.

C

Turn the britch end and roll firmly towards the neck.

D

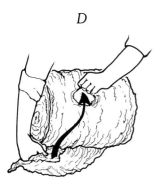

After folding, part the fleece.

E

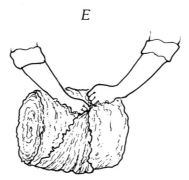

Without twisting, tuck the neck wool firmly into the body of the fleece.

F

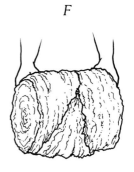

The resulting fleece is firm and secure.

A rolled fleece

Showing butcher's lambs

To show lambs in the butcher's lamb class requires a knowledge of carcase classification. This is a way by which the conformation and degree of fatness is described by visual appraisal, and given a code by the Meat and Livestock Commission staff working in the abattoir. The conformation classification is graded on the scale E,U,R,O,P with E being the best and P the poorest. The fatness is assessed by handling the tail and the spinous and transverse processes of the loin and grading between 1 (the leanest) and 5 (the fattest).

The preferred weight of the carcase is between fifteen and twenty kilograms, so the aim is for a finished carcase grading of 2 or 3 in the conformation class E or U. Grades 3, 4 and 5 are further subdivided into L and H, with L denoting low fat cover and H denoting high. The preferred classification by most meat buyers is within the range of 2 and 3L. The weight of the carcase will be around 45 per cent of the liveweight of the lamb, depending on the degree of fatness.

Great care should be taken at all times with loading and transporting lambs, but particularly if the lamb is to be shown as a carcase. It should never be held or picked up by the wool as this will cause distress to the lamb and result in unsightly bruising of the carcase. Avoid any sharp projections which could cause injury in the transport used to take the lambs to the show, and also avoid abscesses on the carcase caused by faulty injection techniques. Injection sites should be high up on the neck and the hypodermic needles used should be sharp and sterile.

Butcher's lambs

CHAPTER 12 Grassland Management

The flock owner with only a small area of land on which to graze the sheep will perhaps find it difficult to do very much in the way of constructive management to bring about improvement in the grass. Nitrogenous fertiliser used in the early springtime will give a boost to the growth of grass but it must be balanced by adding phosphate and potash at some time in the year or the pasture will become impoverished.

Chain harrowing when the pasture is dry in the early spring and before housed sheep are turned out with their lambs, will remove dead material from the turf and scatter droppings. It will also encourage new growth in the grass.

The pasture can be improved by the clearance of any unpalatable weeds, nettles, thistles, docks, etc. If the acreage is small, the docks can be dug out by hand and the nettles and thistles, if cut regularly through the growing period, will eventually succumb. The sheep will often eat the cut material once it has wilted. Ragwort can pose a problem. The sheep will graze it when it is a rosette of leaves in the early spring without any ill-effects, but later in the season if allowed to grow to the flowering stage it is poisonous to sheep. If it gets included in hay or is cut and left to wilt and then eaten, the accumulation of poison will eventually kill the animal.

Regular mowing to contain the surplus grass even while the sheep are still grazing will improve the pasture. Once the grass gets to the flowering stage, it does not produce much more leaf, which is the nutritious part of the plant. Regular mowing will prevent the grass from flowering or running to seed, thus promoting further growth of leaf.

Regular mowing will also prevent weeds from seeding and increasing their population in the sward. More of the sheep droppings will be exposed to sunlight which will lead to a lower rate of infestation by internal parasites because of the effect it has on the eggs of the roundworms.

If it is possible to subdivide the pasture more involved procedures can be undertaken. The old turf can be treated with a herbicide to kill it off and a rotovator used to cultivate the surface before sowing new grass and clover seed. When choosing new grass seed, select leafy strains of ryegrass for the main part of the mixture and add white clover and deep-rooting herbs. White clover with its ability to fix nitrogen from the air, to the benefit of the accompanying grass, is also high in energy, protein, minerals and vitamins. In addition to giving variety to the mixture, the herbs chicory, burnet and yarrow will help by bringing up minerals from below the level to which the roots of the grass penetrate.

Any quantity of surplus grass can be fenced off using temporary electric fencing and be made into hay. In view of the machinery involved, this is a job for a local contractor. Cut the grass when it is at the early flowering stage and if possible get it dry and baled without allowing any rain on it, then stack it under cover. The regrowth, or aftermath, provides valuable feed for newly weaned lambs as it is clean and the risk of parasitic infection is much less, due to the break in the life cycle of the parasites when the hay crop is taken off. It can be advantageous if cattle are available to assist in controlling the surplus grass and keeping the sward short. This also helps to control parasites.

Subdividing the pasture

CHAPTER 13

Miscellaneous Information Concerning Sheep

SOME PLANTS POISONOUS TO SHEEP

Most poisonous plants are avoided by sheep unless grazing is very sparse as in times of drought or during spells of severe weather, especially snowstorms. The main danger comes when they are cut, wilted and made into hay, often unnoticed, and are then readily eaten by the sheep.

Yew   Probably the plant causing most problems for the small flock owner is the yew. If only a small quantity is eaten the animal will die. This applies to the living shrub or tree, and also to any hedge trimmings which may inadvertently be dumped where the sheep have access to them.

Bracken is poisonous to stock but to a much lesser extent than yew, and will normally be eaten only when grazing is in short supply or when it is made into hay or used as bedding. Sheep affected by bracken poisoning become blind.

Equisetum or horsetail is not usually a problem unless cut, dried and made into hay. If it is then eaten, it is poisonous and there is no antidote. The sheep dies.

Laburnum   Sheep seem to be less affected by eating this tree than some other types of livestock. All parts of the tree are poisonous and will cause death if consumed in quantity.

Acorns are sometimes eaten in quantity by sheep who inexplicably develop a craving for them. They are more poisonous when green and unripe. Affected animals suffer from constipation, which is then quickly followed by diarrhoea. They should be moved away from the source as quicky as possible.

Azaleas and rhododendrons often 'escape' into boundary fences from gardens and woodlands and can cause considerable damage to the nervous system if eaten. As with yew, the clippings from gardens are as toxic when wilted as when they are green. The shrub should either be removed completely or fenced off out of reach of the sheep, as ewes seeking extra nutrition in the weeks just prior to lambing will readily eat it, having avoided it completely for months before.

Ragwort should be pulled out and burned before it reaches the flowering stage (see Chapter 12). The toxin accumulates in sheep if they are given hay containing the dried plant.

SHEEP NAMES

In different parts of the country, different names are given to the various ages of sheep and this can be confusing for the new flock owner. Some names are in common usage, however, and lamb is the universal name given to the newly born. The lambs can be either ram lambs or ewe lambs and they carry this name through the summer even after they are weaned. When the lamb wool breaks to form locks, the wool then forms a tapering point on each lock, similar to the tag on an old-fashioned shoe lace, and tag, or more commonly teg, is the word used to describe this age of sheep. The term hogg or hogget is also used in some parts of the country.

A ewe and her lamb are spoken of as a couple and where there are twins as a double couple.

Chilver is the name given to females in south-west England, and gimmer is commonly used in Scotland and the north of England to describe female sheep after their first shearing. Theave is also a name used in parts of England for once-shorn ewes, while in all areas shearling refers to both males and females between their first and second shearings. Subsequently they can be known as two-shear, three-shear or four-shear rams or ewes.

Sometimes the number of permanent incisor teeth present is used to describe the age of the sheep, and a shearling becomes a two-tooth, a two-shear a four-tooth, a three-shear is a six-tooth and a four-shear is known as full-mouthed.

In hill flocks a substantial income is derived from selling cast ewes. These are ewes which have reached maturity and still have sound teeth and udders, and which can be replaced by the annual intake of shearling ewes bred in the flock. The cast ewes are sold to lowland farmers for crossbreeding where they do not have to withstand the rigours of the climate often experienced by hill sheep. They are sometimes described as draft ewes in other parts of the country. A broken-mouthed ewe or crone is one which has lost some incisor teeth through age but which may still be capable of breeding and rearing lambs, provided that the udder is sound.

A ewe which does not breed when given the opportunity to do so is called either barren, guest or eild.

A cull ewe is one which is not retained for breeding for any reason — chronic ill-health, poor conformation, or as a non-breeder — and which is sold with no guarantee as to its fitness for future breeding.

A wether is a castrated male sheep, and a rig is a male sheep which has either one or both testicles undescended.

SOME ORGANISATIONS SERVING THE SHEEP INDUSTRY

The National Sheep Association (NSA)
The Sheep Centre, Malvern, Worcs WR13 6PH
(01684) 892661
Originally called the National Sheepbreeders Association, it became the National Sheep Association in 1969. It is involved in all aspects of the sheep industry and is financed entirely by members with an interest in sheep and by commercial concerns with a direct interest in the industry. The NSA promotes the industry to those involved in agriculture at home and abroad, and ensures that the appropriate authorities understand the views of the British sheep farmer.

The Rare Breeds Survival Trust
National Agricultural Centre, Kenilworth, Warks CV8 2LG
(01203) 696551

Organic Sheep Society
Worcester College of Agriculture, Hindlip, Worcs WR3 8SS
(01905) 513103

British Wool Marketing Board
Oak Mills, Station Road, Clayton, Bradford, West Yorkshire BD14 6JD
(01274) 882091

National Agricultural Centre
Stoneleigh, Kenilworth, Warks CV8 2LG
(01203) 686503

Meat & Livestock Commission
P.O. Box 44, Winterhill House, Milton
Keynes, Bucks MK6 1AX
(01908) 677577
Royal Agricultural Society of England
National Agricultural Centre, Stoneleigh,
Kenilworth, Warks CV8 2LZ
(01203) 696969

Advice on the design of sheep housing may be obtained from ADAS at the local offices of the Ministry of Agriculture, Fisheries and Food. The address can be found in the local telephone directory.

SHEPHERD'S KIT

Certain essentials of kit and equipment are necessary to enable the beginner to cope adequately with specific tasks which periodically require attention. A suggested list includes:

- *Foot trimming shears* for keeping the sheep's feet in good shape and assisting with footrot prevention.
- *A sharp pocket knife* for paring away under-run horn if infected with footrot.
- *Disposable syringes with spare needles* for vaccinating and administering antibiotics or for injecting calcium borogluconate to counter calcium and magnesium deficiencies.
- *A bottle of calcium borogluconate.*
- *Halters* for holding and restraining sheep when necessary.
- *Dagging shears,* 15cm bladed shears for removing wool soiled with faeces and urine.
- *Sharpening stone.*
- *Combined insecticide and antiseptic,* available in cream or powder form, for treating maggots, shear cuts and external wounds.
- *Antibiotic aerosol spray* for treating wounds, available from your vet.
- *Mild disinfectant, soap and towel.*

- *Disposable plastic gloves,* essential when dealing with sheep infected with the orf virus.
- *Fleece branding irons* for use with marking fluid.
- *Paint aerosols, crayons and tins of marking fluid.*
- *Drenching equipment* for administering anthelmintics.
- *A kneeling pad* for use in inclement weather.
- *Your veterinary surgeon's telephone number.*

These items should be kept ready for use in a clean condition in a suitable container.

SHEEP DAIRYING

Sheep dairying is another option which may be considered by the beginner. The Friesland and the British Milksheep are probably the two main breeds used for milk production, although crosses using these breeds as sires

A milking parlour

105

Friesland and Milksheep

can also give useful yields. A yield of around 250 litres per ewe per year is necessary for a viable enterprise, and further value can be added by converting the milk into yoghurt, cheese or ice cream.

Milk production can be a costly business to set up as, in addition to the cost of the ewes and rams, buildings will be needed to house the equipment and they must be of a high standard. The same standards of hygiene apply as in milk production from cattle. Sheep milk comes under the Food Act (1990) and all areas require inspection by an ADAS officer. Local Environmental Health Officers will wish to inspect areas used for processing, and advice should be sought before starting on such an enterprise.

SHEEP ANNUAL PREMIUM (QUOTA) SCHEME

Should you wish, as a new flock owner, to participate in the Sheep Annual Premium Scheme, it is essential that you obtain details of the scheme as it relates to your holding from the Regional Service Centre of the Ministry of Agriculture, Fisheries and Food before you buy your sheep. You may buy or lease quota from other producers, subject to a number of criteria and only at the discretion of and the approval of an allocation of quota by the Regional Service Centre. The scheme applies to flocks of ten or more eligible ewes or ewe lambs.

MAKING EQUIPMENT

It is possible to save considerable amounts of money by making certain pieces of equipment rather than buying them. An example of this is a hay rack.

Use softwood as sawn. Two 150mm × 25mm planks for lower part of sides, and one 150mm × 25mm for the upper part,

106

nailed or screwed to 75mm × 50mm × 850mm corner posts, with a gap of 350mm between upper and lower parts. Total length not more than 2.5m. Side planks projecting 50mm beyond corner posts at each end, and nailed or screwed to a central upright of 75mm × 50mm × 850mm.

Ends 850mm long with planks nailed to corner posts and finishing flush with them. Four 12mm metal tie rods connecting sides to ends through holes drilled through side planks and corner posts, and secured with nuts and washers.

Use 100mm × 100mm weld mesh cut slightly smaller than the inside area of the hay rack placed on the hay to prevent wastage, or make a 'ladder' from 50mm × 25mm for the same purpose.

If used outside, cover with corrugated iron.

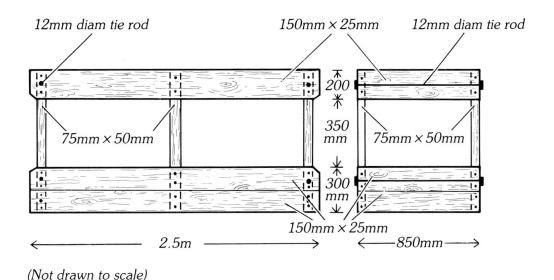

12mm diam tie rod   150mm × 25mm   12mm diam tie rod

75mm × 50mm   200   350 mm   75mm × 50mm   300 mm

150mm × 25mm

2.5m   850mm

(Not drawn to scale)

HAY RACK

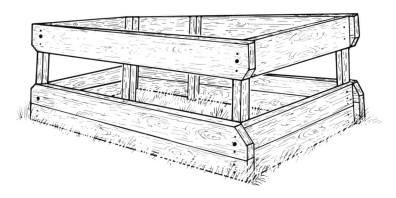

107

SOME SUPPLIERS OF SHEEP EQUIPMENT

Alfred Cox (Surgical) Ltd.
  Edward Road, Coulsden, Surrey
  CR5 2XA
  (0181) 6682131
  A wide range of sheep breeding
  equipment, veterinary instruments and
  hand tools used in caring for sheep.
TWIL Group Marketing Ltd. Fencing
Manufacturers
  P.O. Box 119, Shepcote Lane, Sheffield
  S9 1TY
  Stock fencing and accessories. Free copy
  of Sentinel Fencing Handbook.
Dalton Supplies Ltd.
  Nettlebed, Henley-on-Thames, Oxon
  RG9 5AB
  (01491) 641457
  Ear tags and identification systems and
  livestock equipment.

Gallagher Agricultural Ltd.
  Curriers Close, Canley, Coventry
  CV4 8AW
  (01203) 470141
  Portable and permanent electric fencing.
  Free planning service for electric fencing
  systems.
Ritchey Tagg Ltd.
  Fearby Road, Masham, Ripon,
  N. Yorks. HG4 4ES
  (01765) 689541
  Permanent and temporary sheep
  identification, anti-fighting masks for
  rams, lambing equipment and general
  sheep care products.
Shearwell Ltd.
  Putham Farm, Wheddon Cross,
  Minehead, Somerset TA24 7AS
  (01643) 841552
  Sheep requisites, sheds, handling and
  feeding systems, slatted flooring. Quality
  sheep also available.

IN CONCLUSION

"Sheep are seldom boring and rarely profitable!" to quote a well-known farmer. In spite of this (or because of it?) some twenty million of the creatures are kept in this country. Endearing and infuriating all at the same time, they provide a magnetic interest for those who allow themselves to be attracted to them.

   Treat them well and the rewards will be many.

   Happy sheep keeping!

Index

A

Abortion, 84, 85
Abscesses
    at injection site, 82
    foot, 77, 79
    liver, 55
Acorns, toxicity, 103
ADAS, 105, 106
Adopters, 70-72
Afterbirth, disposal, 49, 59
Animal By-products Order, 89
    transporting animals, 91
Animal Health Division, MAFF, 85
Anthelmintics, 93
    against roundworms and
        tapeworms, 80
    for newborn lambs, 71
Anthrax, 85
Antibiotic spray, 77, 78, 85
Antibiotics at birth, 65
Antiseptic dressing, 75
Auction sales, 3, 50
Azaleas, toxicity, 103

B

Back, conformation, 5
Badgerfaced sheep, 15
Barley straw as bedding, 26
Barren ewes, definition, 104
Battery-operated electric fencer unit, 25
Bedding
    after lambing, 66
    barley straw, 26, 93
Berrichon du Cher breed, 20
Beulah Speckled Face breed, 9
Birth, see Labour
Birth fluids and fostering, 70
Black Welsh Mountain sheep, 15
Blackface breed, 12
Bleu du Maine breed, 19
Bloom dipping, 12
Blowfly strike, 42, 72
    management, 74
    prevention, 75

Bluefaced Leicester breed, 9, 10, 12
Body condition scores, 46-47
    breeding ewes, 48, 49, 50
    rams, 52
Border Leicester breed, 9, 10
Boreray breed, 17
Both front legs back presentation, 63
Bottle feeding, 67
    orphans lambs, 72
Bracken toxicity, 103
Brambles entangling sheep, 22
Branding horns, 86
Brecknock Cheviot breed, 14
Breech or tail first presentation, 64-65
Breeding flock, caring for, 48-53
Breeds from abroad, 19-21
British Milksheep, 105
British Wool Marketing Board, 104
    purchase of fleeces, 44
Broken mouth, 45
Broken-mouthed ewe, definition, 104
Brucellosis, 85
Butcher's lambs, showing, 100

C

Cabbage crops as supplementary food, 37
Calcined magnesite to prevent
    hypomagnesaemia, 83
Calcium borogluconate, 56
    for hypocalcaemia, 83
    for hypomagnesaemia, 83
Calendar, shepherd's, 93-94
Carcases
    classification of lambs, for show, 100
    disposal, 89
Cast ewes, definition, 104
Castration, 49, 68-69
Catching sheep, 30
Cats and abortion, 83
Charollais breed, 19
Cheviot breed, 14
Cheviot Mule, 14
Chilling (hypothermia), newborn lamb, 66
Chilver, definition, 104
Chlamydiosis, 85
Clostridial diseases, 48, 49, 55, 81
    vaccine, 81, 82

Clun Forest breed, 8
Coccidiosis, 85
Colostrum
    amount, 67
    as first feed, 55
    frozen, 55
    making available, 59
    for orphan lambs, 72
    properties, 55
    replacer, 55
    sources, 55
Concentrates, 93, 94
    ewes, flushing, 48
    for supplementary feeding, 36-37
Condition scoring, *see* Body condition
    scores
Conformation, 4-5
Contagious pustular dermatitis (orf), 76-77
    affecting humans, 76
    vaccination against, 77
Contentin breed, 20
Copper sulphate, 85
Cotswold breed, 11
Couple, definition, 104
    double, 104
Crayon markers, 51-52
Crone, definition, 104
Crutching, 42-43
Cull ewes, definition, 104
Culling flocks, 93
    after lambing, 50

D
Dagging, 42
Dagging shears, 105
Dairying, sheep, 105-106
Dalesbred breed, 13
Dartmoor breed, 11
Definitions of sheep terms, 103-104
Dehydration, 80, 85
Derbyshire Gritstone breed, 14
Devon Closewool breed, 11
Diarrhoea, 75, 80, 85
Dipping, 76
    against maggot fly, 45
Diseases, sheep, 74-85
Docking, 49, 69

Dorset Down breed, 6
Dorset Horn breed, 8
Dosing, 40
    guns and syringe, 40
Down breeds, 6
Draft ewes, definition, 104
Dressing fleeces for showing, 98

E
Ears
    notches, 86 87
    tags, 86-87
    tattoo marks, 86
Echinococcus granulosus, 81
Eild ewes, definition, 104
Elastrators, 69
Electrified wire fencing, 23-25
Equipment
    making, 106-107
    sheep, suppliers, 108
Equisetum toxicity, 103
Est a Laine Merino breed, 20
Ewes, breeding
    caring for, 48-50
        after lambing, 49
        body condition scores, 48, 49, 50
        marking lambs, 49-50
        space needed, 49
        supplementary feeding, 49
    culling 50
    vaccinations, 48
Ewes
    guarding against infection, 65
    lambing, 54-66
        see also Labour: Lambing
    per hectare, 2
    unable to regain standing position, 44
Exmoor Horn breed, 14
Eye bolts, 23

F
Face mask for rams, 51
Fatness grading, 100
Feed troughs, cleanliness, 37
Feeding bottles and teats, 55
Feeding troughs, housing, 26
Feet trimming, 50, 93

Fencing, 22-25
  to reinforce hedge, 23-24
Flaked maize, 73
Fleeces
  branding, 88
  dressing for showing, 98
  penalties, 43-44
  preparation for collection, 44
  wool for showing, 98
    rolling a fleece, 99
Flock size and composition, 2, 46
Flooring
  handling pen, 27
  housing, 26
Flushing, 48
Flystrike, *see* Supplementary food
Food Act (1990), 106
Food
  abscesses, 79
  conformation, 5
  trimming, 39
Foot and mouth disease, 85
Footbaths, 29
Footrot, 29, 77-78
  and blowfly strike, 75
  stawberry, 77
  treatment, 78
Formalin, 79
Fostering lambs, 70-72
Friesland sheep, 105
Front leg(s) back presentation, 62

G
Gates for handling pens, 27
Gestation period, 49
Gid, 80
Gimmer, definition, 104
Granulomas, 79
Grass
  control, sheep, 4
  short supply, 36
    supplementary food, 36-38
Grassland management, 101-102
  chain harrowing, 101
  fertilisers, 101
  regular mowing, 101
  subdividing pasture, 101, 102

  surplus grass for hay, 101
Greyface ewe lambs, 12
Greyfaced Dartmoor breed, 11
Guest ewes, definition, 104
Gun for dosing, 40
  paste, 40
  trigger-operated, 40

H
Halter, for rejection of lamb, 56
Halter training, 95-96
Hampshire Down breed, 6
Hand rearing, 72-73
Handling pens, 27-29
  fenced enclosure, 27
  flooring, 27
  gates, 27-28
  housing doubling as, 27
Handling sheep, 30-35
  catching, 30
  immobilising, 30-35
Harness, ram, 48, 51
Hay
  for new lambs, 71
  for orphan lambs, 73
  as supplementary food, 37
Hay making, 101
Hay rack,
  making, 106-107
Head back, both front feet forward
  presentation, 65
Hebridean breed, 17
Hedge as fence, 22, 23-24
Herdwick breed, 13
Hexham breed, 9, 10, 12
Hill breeds, 12-15
Hill Radnor breed, 9
Hind legs first presentation, 63-64
Hogg (hoggett), definition, 104
Horns, branding, 86
Horsetail toxicity, 103
Housing, 26-27
  in bad weather, 93
  doubling as handling pen, 27
  drainage, 26
  draught exclusion, 26
  feeding troughs, 26

flooring, 26
    roof and sides, 26
    space allowance per ewe, 26
    water provision, 26-27
Hydatid disease, 81, 85
Hypocalcaemia, 56, 83
Hypomagnesaemia, 56, 83
    treatment, 83
Hypothermia, newborn lamb, 58

I
Identification, 86-88, 90-91
Immobilising sheep, 30-35
    cradle, 35
    mechanical, 35
    pregnant ewes, 35
    stretcher, 34-35
Incisors,
    and age, 45-46
    inspection after lambing, 50
Infection, guarding against, 65
Injection techniques and site, 100
Inspections of sheep, 44-45
    twice a day, 44
Insulators, 25
Intramuscular injection, vaccine, 41
Iodine, 85

J
Jacob breed, 16
Jaws
    normal, undershot, overshot, 4, 45
Joint-ill, 55

K
Kerry Hill breed, 9

L
Labour
    afterbirth disposal, 59
    assistance at birth, 60
    early signs, 57
    immobilising ewe, 60, 61
    normal presentation, 58-60
Laburnum toxicity, 103
Lamb adopters, 70-72
Lamb creep, 71-72

Lambing, 54-73, 93
    aid, 55, 65
    associated diseases, 85-85
    ewe's behaviour, 56-57
    kit, 54, 55-56
    labour, 57-66
    see also labour
    signs and observations, 55-66
    site, 56-57
Lambs, newborn
    artificial heat, 67-68
    care after birth, 49-50, 58-59, 66-69
    dead at birth, 66
    hand rearing, 72-73
    hypothermia (chilling), 58, 66
    marking, 49
    moving, 58-59
    reviving, 67
    supplementary food, 71
    warming
        by artificial heat, 67-68
        in a heat box, 68
Lambs, supplementary food, 37
Lameness, 77
    in rams, 52
Lanark type of Blackface, 12
Legs, conformation, 5
Listeriosis, 85
Live vaccine against orf, 41
Liver abscess, 55
Liver fluke, 81
Llanwenog breed, 18
Lleyn breed, 9
Lomond breed, 17
Longwool breeds, 9-11
    crutching, 42
Lonk breed, 14
Lubricant, 55

M
Maggot flies, 45, 74, 75
    protection, 45
Magnesium deficiency, 56
Magnesium-fortified calcium borogluconate
    for hypomagnesaemia, 56, 83
Malpresentations, 62-66
Mangolds as supplementary food, 37

112

side effects from shoots, 37
Manx Loghtan breed, 17
Marking crayons, 48-49
Marking Lambs, 71
Masham breed, 13
Masham crossbreed, 10
Mask, ram, 51
Mastitis, 84
 after lambing, 50
Mating, 51-52, 93
Meat & Livestock Commission, 105
Merino breed, 20-21
Metabolic diseases, 82-83
Milk, ensuring supply, 49
Milk fever, 56
Milk production, 105-106
Milk replacer, 67
 for orphan lambs, 72
Mineral supplements, 37-38
Movement of sheep 89, 90
Mules, 10, 12, 13, 14

N
National Agricultural Centre, 104
National Office of Animal Health, 89
National Sheep Association, 3, 104
Navel dressing, 55
Navel infections, 84-85
New Zealand Southdown breed, 21
Nitrogenous fertiliser, 101
 and calcium and magnesium
 deficiency, 83
Norfolk Horn breed, 18
North Country Cheviot breed, 14
North of England Mule, 10, 12, 13
North Ronaldsay breed, 17
Northumberland type of Blackface, 12
Notifiable diseases, 85

O
Oats, 37
One front leg back presentation, 62
Orf (contagious pustular dermatitis), 85, 76-77
 affecting humans, 76
 live vaccine, 41
 vaccination against, 77

Organic Sheep Society, 104
Orphan Lambs, 70-72
 hand rearing, 72-73
Oxford Down breed, 7

P
Pasteurella pneumonia (pasteurellosis), 82
 vaccination, 82
Pasture, see Grassland management
Pellets, Lamb, 37
Perth type of Blackface, 12
Pessaries, 55
Pneumonia, 82
Poisonous plants, 22
Pole barn construction, 26
Polled Dorset breed, 8
Polwarth breed, 20
Portland breed, 17
Pregnancy toxaemia, 82-83
Pregnant ewes
 immobilising 35
 supplementary food, 36-37
Pregnant women and sheep abortion, 84
Presentations, see Labour: Malpresentations:
 Twin lambs
Prolapse, vaginal, 56
Protein, supplementary pregnant ewe, 36-
 37
Proudflesh, 79
Pulpy kidney, 81-82

Q
Quota scheme, 106

R
Rabies, 85
Radisseurs, 23
Ragwort toxicity, 101, 103
Rams
 harness, 48
 mating, 51-52, 93
 records, 88
 selection and care, 50-53
 body condition score, 52
 crayon markers, 51-52
 face mask, 51

harness, 51
hiring or purchasing, 53
   infection risks, 53
lameness, 52
numbers of ewes served, 52
selection, 50-51
Rare breeds, 17-18
Rare Breeds Survival Trust, 3, 104
Record Keeping, 86-92
   disposal of carcases, 89
   ewe record card, 88
   health records, 89
   identification, 86-88, 90-91
   medicinal, 89
   movement, 89 90
   ram, 8
   registration, 90
   Sheep and Goat Registration,
   Identification
      and Movement Order 1996, 89-90
   Sheep Annual Premium Scheme, 89,
      90, 106
   systems, 88-91
   cards, 88
      loose-leaf, 88
   transport, 91
      legal requirements, 91
Registration of sheep, 90
Rhododendrons, toxicity, 22, 103
Rib, conformation, 5
Ringwomb, 60
Romney breed, 9
Root vegetables as supplementary food, 37
Rope, leading, 96
Rouge de l'Ouest 20
Rough Fell breed, 13
Roundworms, 79-80
Routine tasks, 39-46
Royal Agricultural society of England, 105
Ryeland breed, 8

S
Sale sheep, supplementary food, 37
Salmonellosis, 85
Scab, sheep, 76
Scald, 29 77-78
   treatment, 78

Scores for body conditioning, 46-47
Scottish Halfbred, 14
Scrotal hernia, 69
Selection of sheep
   rams, 50-51
   for showing, 95
Shearing, 43-44
   12-hour food deprivation, 43
   dry wool, 43
   registration, 44
   skin cuts, 44
Shearling, definition, 104
Sheep Annual Premium (Quota) Scheme,
   89, 91, 106
Sheep auctions, 3
   ewes after lambing, 50
Sheep and Goat Registration, Identification
   and Movement Order 1996, 89-90
Sheep hurdles, 28-29
Sheep pox, 85
Sheep scab, 76
Sheep subsidies, records, 91
Sheep/shepherd's clubs, 2
Shepherd's calendar, 93-94
Shepherd's kit, 105
Shetland breed, 16-17
Shortwool breeds, 6-9
Show sheep, supplementary food, 37
Showing, 95-100
   butcher's lambs, 100
   dressing, 98
   fleece wool, 98
   halter training, 95-96
   leading on long rope, 96
   selection, 95
Showmanship, 97
Shropshire breed, 8
Size of flock, 2
Skinning lamb before fostering, 70
Soay breed, 18
Southdown breed, 6
Space allowance, housing, 26
Standing correctly for showing, 96
Stock markers, 56
Stocking rates, 2
Stomach tube with syringe, 55
Strainer posts, 23

Straw. 26, 93
Strawberry footrot, 77
Stretcher, sheep 34-35
Strike, *see* Blowfly strike
Subcutaneous injection, vaccine, 41
Subsidies, records, 91
Suckling by lamb, reluctance, 67
Sudden death, causes, 81-81
Suffolk breed, 8
Sugar beet pulp as supplementary food, 37
Supplementary food, 36-38
   lambs, newborn, 71
   pregnant ewes, 49
Swaledale breed, 12
Syringe for dosing 40
   for colostrum administration, 55, 67

T
Tag, definition, 104
Tapeworms, 80
Tattoos, 86
Teeswater breed, 9, 10
Teeth, 45-46
   incisors and age, 45
     inspection before breeding, 50
   see also Jaws
Teg, definition, 104
Tensioners for fencing, 23
Testicles, undescended, 69, 104
Texel breed, 19
Theave, definition, 104
Torddu sheep, 15
Torwen sheep, 15
Toxaemia, pregnancy, 82-83
Toxoplasmosis, 85
Trailers, 91, 92
Transport, record keeping, 91
   legal requirements, 91
Trough space, adequate, 37
Twin lamb disease, 82-83
Twin lambs
   one rejected by ewe, 71
   presented together, 65

U
Udder inspection before breeding, 50

Umbilical cord
   dressing, 66
   and infection, 84-85

V
Vaccination, 41, 81-82, 93, 94
   administration, 41
   for all bought-in sheep, 81-82
   before mating, 48
Vaginal prolapse, 56
   support, 56
Vendeen breed, 20
Vetrazin for blowfly strike, 75

W
Water containers and troughs, cleanliness, 37
Water provision, housing, 26
Waterbag, 57
Weighing lambs, 72
Welfare of Animals During Transport Order 1994, 89
Welfare of Animals Transport Order 1997, 89
Welsh Halfbred, 14
Welsh Mountain breeds, 14-15
Welsh Mule, 14
Wensleydale breed, 9, 10
Wether lambs for grass control, 4
   definition, 104
Whitefaced Dartmoor breed, 11
Wiltshire Horn breed, 16
Winders, 23
Wire fence, 22-25
Wool, 37
   for showing, 98
   see also Fleeces
Worms, dosing, 93

Y
Yew toxicity, 22, 103
Yorkshire cladding, 26

Z
Zinc sulphate for footrot or scald, 29, 78
Zoonoses, 85

FARMING PRESS BOOK & VIDEOS

Below is a sample of the wide range of agricultural and veterinary books
and videos we publish.
For more information or for a free illustrated catalogue
of all our publications please contact:

**Farming Press
Miller Freeman UK Ltd
2 Wharfedale Road, Ipswich IP1 4LG, United Kingdom
Tel (01473) 241122 Fax (01473) 242222 E-mail: farmingpress@unmf.com**

The Veterinary Book for Sheep Farmers

DAVID C. HENDERSON

A wide-ranging, detailed guide to the prevention of sheep ailments, to increased lamb output and to the diagnosis and treatment of disease.

Sheep Ailments

EDDIE STRAITON

All the essential information that should be at the fingertips of the shepherd, flock owner or student. Some 400 photographs and drawings, almost half in colour.

Genetic Improvement of Cattle and Sheep

GEOFF SIMM

A fascinating and practical introduction to genetics for the farmer and student. The basic principles of genetics and the application of these principles to today's livestock farming are clearly set out.

Lamb Survival (VHS video)

DAVID C. HENDERSON

Full discussion of the techniques to keep losses from hypothermia and infection to a minimum.

Wattle Hurdles and Leather Gaiters

JOHN RANDALL

Detailed recollections of working with sheep and other livestock from the 1920s onwards, told with humour and affection by a quintessential countryman.

Showman Shepherd

DAVID TURNER

A guide to all aspects of showing sheep.

All about Goats

LOIS HETHERINGTON

The ideal introduction to keeping goats.

Farming Press is a division of Miller Freeman UK Ltd
which provides a wide range of media services in agriculture and
allied businesses. Among the magazines published by the group are:
**Arable Farming,
Dairy Farmer,
Farming News**
and **Pig Farming.**
For a specimen copy of any of these please contact the address above.